Python for Scientific Computing and Artificial Intelligence

Python for Scientific Computing and Artificial Intelligence is split into 3 parts: in Section 1, the reader is introduced to the Python programming language and shown how Python can aid in the understanding of advanced High School Mathematics. In Section 2, the reader is shown how Python can be used to solve real-world problems from a broad range of scientific disciplines. Finally, in Section 3, the reader is introduced to neural networks and shown how TensorFlow (written in Python) can be used to solve a large array of problems in Artificial Intelligence (AI).

This book was developed from a series of national and international workshops that the author has been delivering for over twenty years. The book is beginner friendly and has a strong practical emphasis on programming and computational modeling.

Features:

- No prior experience of programming is required.
- Online GitHub repository available with codes for readers to practice.
- Covers applications and examples from biology, chemistry, computer science, data science, electrical and mechanical engineering, economics, mathematics, physics, statistics and binary oscillator computing.
- Full solutions to exercises are available as Jupyter notebooks on the Web.

In 2022, Stephen Lynch was named a National Teaching Fellow, which celebrates and recognises individuals who have made an outstanding impact on student outcomes and teaching in higher education. He won the award for his work in programming in the STEM subjects, research feeding into teaching, and widening participation (using experiential and object-based learning). Although educated as a pure mathematician, Stephen's many interests now include applied mathematics, cell biology, electrical engineering, computing, neural networks, nonlinear optics and binary oscillator computing, which he co-invented with a colleague. He has authored 2 international patents for inventions, 8 books, 4 book chapters, over 40 journal articles, and a few conference proceedings. Stephen is a Fellow of the Institute of Mathematics and Its Applications (FIMA) and a Senior Fellow of the Higher Education Academy (SFHEA). He is currently a Reader with MMU and was an Associate Lecturer with the Open University from 2008-2012. In 2010, Stephen volunteered as a STEM Ambassador, in 2012, he was awarded MMU Public Engagement Champion status, and in 2014 he became a Speaker for Schools. He runs national workshops on "Python for A-Level Mathematics and Beyond," and international workshops on "Python for Scientific Computing and TensorFlow for Artificial Intelligence." He has run workshops in China, Malaysia, Singapore, and the USA.

Chapman & Hall/CRC
The Python Series

About the Series

Python has been ranked as the most popular programming language, and it is widely used in education and industry. This book series will offer a wide range of books on Python for students and professionals. Titles in the series will help users learn the language at an introductory and advanced level, and explore its many applications in data science, AI, and machine learning. Series titles can also be supplemented with Jupyter notebooks.

Image Processing and Acquisition using Python, Second Edition
Ravishankar Chityala, Sridevi Pudipeddi

Python Packages
Tomas Beuzen and Tiffany-Anne Timbers

Statistics and Data Visualisation with Python
Jesús Rogel-Salazar

Introduction to Python for Humanists
William J.B. Mattingly

Python for Scientific Computing and Artificial Intelligence
Stephen Lynch

For more information about this series please visit: https://www.crcpress.com/Chapman--HallCRC/book-series/PYTH

Python for Scientific Computing and Artificial Intelligence

Stephen Lynch

Manchester Metropolitan University, United Kingdom

CRC Press
Taylor & Francis Group
Boca Raton London New York

CRC Press is an imprint of the
Taylor & Francis Group, an **informa** business
A CHAPMAN & HALL BOOK

First edition published 2023
by CRC Press
6000 Broken Sound Parkway NW, Suite 300, Boca Raton, FL 33487-2742

and by CRC Press
4 Park Square, Milton Park, Abingdon, Oxon, OX14 4RN

CRC Press is an imprint of Taylor & Francis Group, LLC

© 2023 Stephen Lynch

Library of Congress Cataloging-in-Publication Data

Names: Lynch, Stephen, 1964- author.
Title: Python for Scientific Computing and Artificial Intelligence /
Stephen Lynch.
Description: First edition. | Boca Raton : C&H/CRC Press, 2023. | Series:
Chapman & Hall/CRC the Python series | Includes bibliographical
references and index.
Identifiers: LCCN 2022057798 (print) | LCCN 2022057799 (ebook) | ISBN
9781032258737 (hardback) | ISBN 9781032258713 (paperback) | ISBN
9781003285816 (ebook)
Subjects: LCSH: Python (Computer program language) | Science--Data
processing. | Artificial intelligence.
Classification: LCC QA76.73.P98 L96 2023 (print) | LCC QA76.73.P98
(ebook) | DDC 005.13/3--dc23/eng/20230125
LC record available at https://lccn.loc.gov/2022057798
LC ebook record available at https://lccn.loc.gov/2022057799

ISBN: 978-1-032-25873-7 (hbk)
ISBN: 978-1-032-25871-3 (pbk)
ISBN: 978-1-003-28581-6 (ebk)

DOI: 10.1201/9781003285816

Typeset in LM Roman
by KnowledgeWorks Global Ltd.

Publisher's note: This book has been prepared from camera-ready copy provided by the authors

To my family,
my brother Mark and my sister Jacqueline,
my wife Gaynor,
and our children, Sebastian and Thalia,
for their continuing love, inspiration and support.

Contents

Foreword

I came to know about the author through his previous text: *Dynamical Systems with Applications using Python*, Springer International Publishing (2018). I simply loved this book as it helped me to enhance my knowledge in both Python and Dynamical Systems. I was so pleased with this book that in collaboration with a senior colleague in Dynamical Systems, we developed a set of two-semester courses with the same title as the book. We also invited Stephen to offer for us a short course on Python for Scientific Computing and TensorFlow for Artificial Intelligence (AI). We were very pleased that he accepted our offer, as he has been in demand to offer these courses all over the world. For Pitt, he first offered it online (due to Covid) in summer 2021, and then in person on Pitt campus in summer 2022. Both of these courses were huge successes with several hundreds of participants each time. These delegates came from a wide variety of backgrounds: professors, scientists and engineers from a wide range of disciplines in industry & government labs, and both graduate and undergraduate students in a broad range of disciplines. The excellent textbook plus Stephen's natural teaching talents made the course very popular. We plan to offer it on a regular basis as long as Stephen is up for it.

Based on these experiences, I was naturally very eager and excited to read Stephen's newest (present) text, and I am not surprised to see that this volume is also a masterpiece! The topics and approach taken in this book are indeed like no other! The first section provides an introduction to Python for complete novices and goes on to show how it can be used for teaching High School mathematics. Readers are shown how to access Python through the IDLE platform, Spyder and jupyter notebooks through Anaconda, and cloud computing using Google Colab. The second section covers a vast array of examples from the scientific fields of biology, chemistry, data science, economics, engineering, fractals & multifractals, image processing, numerical methods for ordinary and partial differential equations, physics and statistics. Each chapter covers several examples from the field and there are exercises with even more examples at the end of each chapter. The final section covers AI, with brain-inspired computing and neurodynamics and neural networks following on nicely from Section 2. The final chapters introduce TensorFlow and Keras, and recurrent and convolutional neural networks, respectively. There is further reading for those interested in cyber security, ethics in AI, the internet of things, natural language processing and reinforcement learning.

All Python and TensorFlow programs can be downloaded through GitHub and full worked solutions for all of the exercises are available to view on the Web. Overall, the book will appeal to a very wide range of readers, from high school, to undergraduate and postgraduate students in science & engineering, to academics and researchers in industry. Novices to programming and even experienced programmers will all get something from this book.

Professor Lynch has recently been named a **National Teaching Fellow** in the UK. I am not very familiar with UK's system. But as I understand, this fellowship is "...to recognize and celebrate individuals who have made an outstanding impact on student outcomes and the teaching profession..." Based on his other books, and his courses, I can easily see that Stephen is surely deserving of this recognition. He is an excellent writer, with an exceptional knack for teaching.

I hope that the reader is as excited as I was to get started on this book.

Professor Givi Peyman, Distinguished Professor of Mechanical Engineering and the James T. MacLeod Professor in the Swanson School of Engineering at the University of Pittsburgh.

Preface

This book has developed from a series of university, national and international workshops that I have been delivering for over twenty years. It is based on the programming language Python, which is a high-level programming language and supports functional, imperative, object-oriented and procedural styles of programming. Probably the most important aspect of Python is that it is open source and completely free to the user. Python is currently the most popular programming language in the world from over 600 programming languages, and it is a relatively user-friendly language. The book is split into three sections. In Section 1, the reader is introduced to the Python programming language and it is shown how Python can aid in the understanding of A-Level (or High School) Mathematics. In Section 2, the reader is shown how Python can be used to solve real-world problems from a broad range of scientific disciplines and in Section 3, the reader is introduced to neural networks and shown how TensorFlow (written in Python) can be used to solve a large array of problems in Artificial Intelligence (AI). As well as learning the most popular programming language in the world, the reader will also be shown how Python can be used to perform computational mathematics to solve real-world problems. By the end of the book, the reader will be able to solve a great many mathematical problems without having to learn any of the theory of mathematics. The reader should note that the emphasis of the book is on programming and computational modeling and not on the mathematical theory. For those readers who want to know all of the ins-and-outs of the mathematics—for High School Mathematics, there are recommended books listed in Section 1, and for Sections 2 and 3, I recommend my other book, Dynamical Systems with Applications using Python [1], which covers the mathematical theory in some detail. To keep the book relatively brief, explanations of the syntax of the code are not included. Instead, the reader is given full working programs and explanations for syntax are easily discovered on the Web.

Chapter 1 starts with the Python Integrated Development Learning Environment (IDLE) which can be downloaded at:

https://www.python.org.

In Section 1.1, the reader is introduced to the IPython Console Window, where users can use Python as a powerful calculator. As every reader is already familiar with a calculator, I have found this method of teaching the best way to introduce Python. The Math library (or module) is introduced and the functions available can be found using the Help features in IDLE. Next, the reader is shown how to define their own functions in Python—think of adding buttons to your calculator. For and while loops are introduced by means of simple examples and then if, elif, else constructs are discussed. With these three programming structures, one can write any of the programs listed throughout the book! To further aid in the understanding of how programs work, the turtle module is introduced in Section 1.3. Readers may have heard of the Scratch programming language:

`https://scratch.mit.edu`

developed at MIT, which is the largest free coding community for kids in the world. Well, the turtle module is similar to Scratch and provides a pictorial representation to coding and that is why it is so popular with children. In this book, I use turtle to plot simple colorful fractals using iteration and readers can figure out how the programs work by slowing down the turtle, plotting early stage fractals, and following the code by drawing little pictures—this helps greatly when learning to code. Chapter 1 ends with some exercises and a short bibliography, where interested readers can find more introductory texts for Python. It is highly recommended that the reader attempts the exercises at the end of Chapter 1 before moving on to Chapter 2. Note that, solutions to the exercises, as in all other chapters can be viewed on the Web, the corresponding URL will appear at the end of each chapter. IDLE is not suitable for scientific computation, for that you will need Anaconda or Google Colab. In Chapter 2, the reader is shown how to download Anaconda, the world's most popular data science platform:

`https://www.anaconda.com/products/individual,`

simply download Anaconda installers for Linux, Mac or Windows. Within this framework, we can launch Spyder (Scientific Python Development Environment) or a jupyter notebook. Those of you familiar with MATLAB®, will see similarities with Spyder and if any readers have ever used Mathematica®, you will notice a similarity with both sets of notebooks. Note that within Anaconda there are other tools available including the R studio for statistics and Orange 3 and Glueviv for data visualization and data analysis. In this book, I will only be using Spyder and jupyter notebooks from Anaconda. Note that although Anaconda is well known for Data Science, it is much more flexible than that, and is ideal for both scientific computing and AI. Section 2.1 is concerned with numerical Python using the numpy library (or module):

`https://numpy.org,`

where lists, arrays, vectors, matrices and tensors are introduced. These concepts are vitally important in neural networks, machine learning and deep learning, which will all be covered in the third section of the book. The chapter ends with a tutorial introduction to the matrix plotting library matplotlib:

`https://matplotlib.org,`

used for creating static, animated and interactive visualizations in Python. As well as producing colorful plots, readers will be shown how to save their graphs to file in any format they like (eps, jpeg, png, tiff etc.) and change the resolution of figures using dpi (dots per inch). They will also be shown how to label axes and figures with LaTeX symbols:

`https://www.caam.rice.edu/~heinken/latex/symbols.pdf.`

In Section 2.2, the symbolic library (or module) sympy is introduced:

`https://www.sympy.org/,`

by means of example. Sympy is a powerful symbolic computation library for Python and it enables the reader to perform some complex mathematics. Here the reader will be shown how to differentiate and integrate without needing to know all of the mathematical rules. Matrices and simple matrix algebra will be covered here.

Chapter 3 provides an introduction to both jupyter notebooks, accessed through Anaconda, and Google Colab jupyter notebooks, which are accessed through cloud computing. When you download Anaconda onto your machine you take up memory on your computer. The major benefit with this method is that you do not need access to the Web to run your Python code when using Spyder. Google Colab, on the other hand, is cloud computing, where your computations are performed through Google on the cloud:

`https://colab.research.google.com/.`

Note that you need a Google account in order to have access to Google Colab. The major benefit with this method is that you do not use up memory on your own computer when you run the Python code. Section 3.1 shows the jupyter notebook interface and readers are shown how to use Markdown to insert titles, text, html code and LaTeX mathematical equations into a notebook. Code cells are used to insert Python code. Cells are compiled by hitting SHIFT + ENTER in the cell or by running the cell using the small play button. In Section 3.2, the reader will be shown how to load files and figures into a notebook. Examples demonstrating animations and interactive plots will be listed here. The notebook can be saved as a Python notebook (filename.ipynb), a Python file (filename.py) or as a html file (filename.html) so the notebook can be published on the web. In Section 3.3, Google Colab and GitHub will be introduced and this again will be useful material for what is to come in the third section of the book. Chapters 4–5 show how Python can be used to give a deeper understanding of the new A-Level Mathematics syllabus which was introduced to the UK in 2017. Most of the topics covered here are covered by many high schools around the world and once more it is demonstrated how Python can be used to solve mathematical problems without understanding all of the mathematical theorems and rules. A-Level (and High School) pupils will find these chapters extremely useful for providing a deeper understanding of the mathematics.

Section 2 of the book covers scientific computation and follows on nicely from Section 1. Chapter 6 presents examples from the field of Biology starting with a simple single population model of insects using the logistic model and iteration. Fixed points, stability and bifurcation diagrams are introduced. Phase plane portraits are plotted in the next section and the behavior of interacting species is discussed. The spread of flu in a school using a compartmental model is covered next. In Section 6.4, the reader can reproduce results from a recent journal paper where a spring-damper model is used to simulate hysteresis (the system is history dependent) in single fiber muscle. Chapter 7 covers Chemistry, balancing chemical-reaction equations, solving differential equations to model chemical kinetics, and investigating the solubility of silver chloride in a potassium chloride solution. Chapter 8 covers some topics from the popular field of data science. The Pandas package is introduced for creating data frames for data manipulation and analysis. Linear programming is introduced to solve optimization problems, including the simplex method. Unsupervised learning is used to cluster data in the third section, and the final section covers decision trees. The field of data science is huge, and readers should note that statistics, machine learning, and deep learning, covered later in the book, are also encompassed within data science. In Chapter 9, the reader is shown a number of examples from the field of Economics, including a microeconomic model of quantity of production, a macroeconomic model of economic growth, the Markowitz model—a portfolio optimization model and the Black-Scholes model for computing the fair price or theoretical value for a call or a put option based on a function of four variables. In the field of engineering, linear and nonlinear electric circuits (including Chua's circuit) are covered in the first half of Chapter 10, and then mechanical systems of coupled oscillators and nonlinear pendula are modeled. Chapter 11 is concerned with fractals, multifractals and real-world applications. Image processing is covered in Chapter 12, both grayscale and color images are investigated, and methods for finding features, edge detection and statistical analysis of images are covered. The chapter ends with examples from medical imaging, plotting the vascular architecture from a human retinal image using a ridge filter, and identifying a tumor in a brain image using masking and segmentation. In the first half of Chapter 13, numerical methods for Ordinary Differential Equations (ODEs), including Euler's methods, the fourth order Runga-Kutta method and the Adams-Bashforth method, are all introduced. Numerical methods for Partial Differential Equations (PDEs), and in particular finite difference methods, are covered in the second half of Chapter 13. The standard examples of heat diffusion in a rod and a flat plate, advection along a one-dimensional channel, and the the vibration of a string are all included. Chapter 14 is concerned with the field of Physics, and examples are taken from signal processing, using Fast Fourier Transforms (FFTs) and nonlinear optics, where the Ikeda map is used to model the propagation of light waves through a Simple Fiber Ring (SFR) resonator. Josephson junctions, superconducting devices that act like biological neurones but oscillate 100,000,000 times faster are discussed, and the motion of planetary bodies (earth-moon-comet system) is also covered. Chapter 15 shows how Python can be used for statistics, including linear regression, the student t-test, Markov chains and Monte-Carlo simulation, for example.

The final section of the book is concerned with neural networks and AI. To follow Section 2, scientific computation is applied to modeling neurones and simple neural networks in some detail. Machine learning, deep learning and TensorFlow (written in Python) are covered later. Chapter 16 covers an invention by the author (and one of his colleagues) labeled brain inspired computing. The first section deals with the Hodgkin-Huxley model from the 1950s before moving on to later models that have been developed over the decades. Using threshold oscillator logic, a binary oscillator half-adder for logic computation is simulated in Section 16.2, and in the next section a memory component is modeled using an oscillator-based set reset flip-flop. The chapter ends with applications and future work. Chapter 17 covers neural networks and neurodynamics. Simple neural networks are used to model logic AND, OR and XOR gates. Section 17.2 covers the backpropagation algorithm, with a simple explanation of how this works, and then applied to a neural network model for the Boston Housing data for valuing homes. The final section presents a simple model of a two-neuron module. A stability analysis is carried out to determine parameter borders where the system displays hysteresis, unstable behavior and quasiperiodic (almost periodic) behavior. Chapter 18 provides an introduction to TensorFlow and the Keras Application Programming Interface (API). TensorFlow is then used to build networks for linear regression, the XOR gate and finally the Boston Housing data valuations. Recurrent Neural Networks (RNNs) are covered in Chapter 19, starting with discrete and continuous Hopfield models used as associative memory RNNs. The long short-term memory RNN network is applied to predicting chaotic and financial time series in sections three and four. Chapter 20 presents an example of a convolutional neural network (CNN). A simple example illustrating convolving and pooling starts the chapter, and then a CNN is used on the MNIST handwritten dataset for predicting the digits 0-9. In the next section, TensorBoard, which is TensorFlow's visualization toolkit, is used to see how well your neural network is performing. The final section covers further reading, natural language processing (NLP), reinforcement learning, ethics in AI, cyber security and the internet of things (IoT) are briefly discussed. Chapter 21 presents answers or hints to all of the exercises in the book.

Full solutions to the Exercises in Sections 1, 2 and 3, may be found in the URLs below:

https://drstephenlynch.github.io/webpages/Solutions_Section_1.html.

https://drstephenlynch.github.io/webpages/Solutions_Section_2.html.

https://drstephenlynch.github.io/webpages/Solutions_Section_3.html.

Finally, all of the IDLE, Python and jupyter notebooks can be dowloaded through GitHub:

https://github.com/proflynch/CRC-Press/.

[1] Lynch, S. (2018). *Dynamical Systems with Applications using Python*. Springer International Publishing, New York.

I

An Introduction to Python

The IDLE Integrated Development Learning Environment

Python is currently the most popular programming language in the world; it is open source and freely available, which makes it the ideal package to support the learning of A-Level (High-School) Mathematics, to solve real-world problems in Scientific Computing and to grapple with Artificial Intelligence. Python is a high-level language developed by Guido van Rossum and first released in 1991. Python 2.0 was released in 2000 but was discontinued as version 2.7.18 in 2020. Python 3.0 was first released in 2008, and the latest version of Python, version 3.11.0 was released in October 2022.

Note that all of the Python programs that accompany the book will be available through GitHub, where readers can download all Python and notebook files. This means that the book should not go out of date, as any changes in Python can be relayed using those media.

In this chapter, the reader is first shown how to use Python as a powerful calculator—all readers should have used a calculator at school and I have found this approach to be popular with my students. Next, simple programming constructs are discussed. Functions are defined—the reader can think of this as adding buttons to your Python calculator. For and while loops are covered next by means of simple examples and if, then, else constructs round off the section. Finally, the reader is shown how to plot simple fractals (images repeated on ever reduced scales) using the turtle module. In this way, readers are able to understand programming constructs by plotting simple pictures.

The chapter ends with some simple exercises and a short bibliography for further reading.

DOI: 10.1201/9781003285816-1

Aims and Objectives:

- To introduce programming with Python.

- To provide tutorial guides to the Python modules.

- To provide a concise source of reference for experienced users.

At the end of the chapter, the reader should be able to

- use Python as a powerful calculator;

- define functions and construct simple programs incorporating loops and conditional statements;

- construct simple fractals using the turtle library (module).

1.1 INTRODUCTION

Figure 1.1 shows the homepage for Python, and readers can download the latest version of Python by hitting the Download tab (available at the bottom of the Web page). One can download on a Mac, Windows or Linux computer. The reader is first introduced to IDLE (an Integrated Development Learning Environment) which can be downloaded here:

`https://www.python.org.`

The icon to launch Python IDLE is shown on the left-hand side of Figure 1.2. Readers can use either a shell window or an editor window.

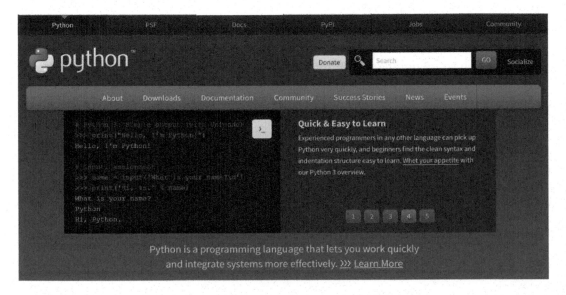

Figure 1.1   The official Python programming homepage.

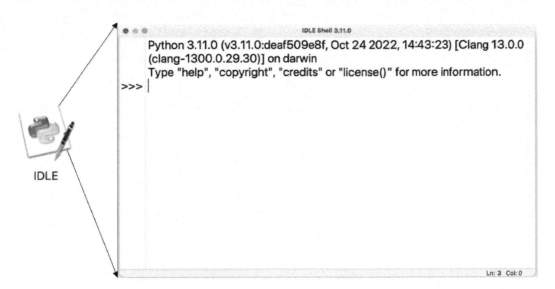

Figure 1.2 The IDLE Shell Window as it first appears when opened on a Macbook. The chevron prompt >>> indicates that Python is waiting for input. The IDLE shortcut icon is shown to the left of the IDLE Shell Window.

1.1.1 Tutorial One: Using Python as a Powerful Calculator (30 Minutes)

The reader should copy the commands after the chevron prompt >>> in the IDLE Shell (see Figure 1.2). There is no need to copy the comments; they are there to help you. Note that gaps are left between operators in Python; this is simply to make Python code look pretty. The output is not included here but can be seen in the Jupyter notebook for this chapter. Note that help(math) will list all of the functions in the math library. Essentially, this is showing you the buttons (functions) on your powerful Python calculator. Later in this chapter, I will show you how to define your own functions (buttons) in Python.

Python Command Lines	Comments
>>> # This is a comment.	# Writing comments in Python.
>>> 4 + 5 - 3	# Addition and subtraction.
>>> 2 * 3 / 6	# Multiplication and division.
>>> 2**8	# Powers.
>>> import math	# Import the math library (or module).
>>> help(math)	# List the functions.
>>> math.sqrt(9)	# Prefix math for square root.
>>> from math import *	# Import all math functions.
>>> sin(0.5)	# The sine function (radians).
>>> asin(0.4794)	# Inverse sine function.

```
>>> degrees(pi)                    # Convert radians to degrees.

>>> radians(90)                    # Convert degrees to radians.

>>> log(2)                         # Natural logarithm.

>>> log10(10)                      # Logarithm base 10.

>>> exp(2)                         # Exponential function.

>>> e**2                           # Exponential function using e.

>>> cosh(0.3)                      # Hyperbolic coshine function.

>>> fmod(13, 6)                    # Modulo arithmetic.

>>> 13 % 6                         # Returns the remainder.

>>> gcd(123, 321)                  # Greatest common divisor.

>>> 1 / 3 + 1 / 4                  # Floating point arithmetic.

>>> from fractions import Fraction # Load the fractions function Fracti

>>> Fraction(1, 3) + Fraction(1, 4) # Symbolic computation.

>>> pi                             # The number π.

>>> round(_, 5)                    # Round last output to 5 decimal pla

>>> factorial(52)                  # Gives 52!

>>> ceil(2.5)                      # Ceiling function.

>>> floor(2.5)                     # Floor function.

>>> trunc(-2.5)                    # Truncates nearest integral to zero

>>> quit()                         # Quit Python IDLE.
```

Most of the functions from the math module will be familiar to most readers, however, the **factorial, ceiling, floor** and **truncate** functions are worthy of further discussion.

The number $52! = 52 \times 51 \times 50 \times \ldots \times 3 \times 2 \times 1$ is equal to:

$$52! = 30414093201713378043612608166064768844377641568960512000000000000,$$

and this is the number of permutations of a normal pack of playing cards. Therefore, if you give a deck of cards a really good shuffle, the order of the cards is probably unique. Note that no normal calculator can cope with this number.

The functions **ceiling, floor** and **truncate** are probably new to most readers. The function **ceil(n)** gives the smallest integer greater than or equal to n, where n is a real number. The function **floor(n)** gives the largest integer less than or equal to n, where n is a real number. The function **truncate(n)** gives the nearest integer to zero on the real line. The reader is encouraged to try their own examples in order to get a deeper understanding.

1.1.2 Tutorial Two: Lists (20 Minutes)

A list is the most versatile data type in all functional programming languages. Later in the book, the reader will be introduced to arrays, vectors, matrices and tensors. A list is simply a tool for storing data. In this tutorial, readers will be shown how to define lists and how to access the elements in a list.

Python Command Lines	Comments
`>>> a = [1, 2, 3, 4, 5]`	`# A simple list.`
`>>> type(a)`	`# a is a class list.`
`>>> a[0]`	`# 1st element, zero-based indexing.`
`>>> a[-1]`	`# The last element.`
`>>> len(a)`	`# The number of elements.`
`>>> min(a)`	`# The smallest element.`
`>>> max(a)`	`# The largest element.`
`>>> 5 in a`	`# True, 5 is in a.`
`>>> 2 * a`	`# [1,2,3,4,5,1,2,3,4,5].`
`>>> a.append(6)`	`# Now a=[1, 2, 3, 4, 5, 6]`
`>>> a.remove(6)`	`# Removes the first 6.`
`>>> print(a)`	`# a=[1, 2, 3, 4, 5].`
`>>> a[1 : 3]`	`# Slice to get [2, 3].`
`>>> list(range(5))`	`# [0, 1, 2, 3, 4].`
`>>> list(range(4 , 9))`	`# [4, 5, 6, 7, 8].`
`>>> list(range(2, 10, 2))`	`# [2, 4, 6, 8].`
`>>> list(range(10, 5, -2))`	`# [10, 8, 6].`
`>>> A = [[1, 2], [3, 4]]`	`# A list of lists.`
`>>> A[0][1]`	`# Second element in list one.`
`>>> names = ["Jon", "Seb", "Liz"]`	`# A list of names.`
`>>> names.index("Seb")`	`# Returns 1.`
`>>> names.pop(1)`	`# Returns 'Seb' and removes from names.`
`>>> quit()`	`# Quits Python IDLE.`

Note that Python uses zero-based indexing so that the first item has index zero, and the command **range(n)** starts with zero and lists the integers up to but not including n.

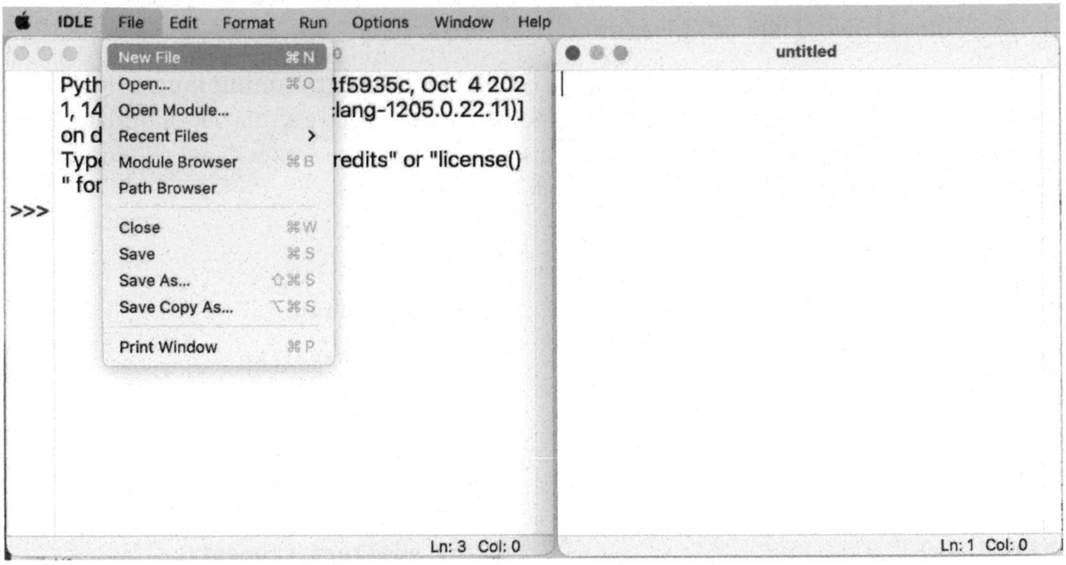

Figure 1.3 From IDLE click on **File** and **New File** to get the untitled Editor Window on the right.

1.2 SIMPLE PROGRAMMING IN PYTHON

There are plenty of published introductory books on Python; the reader may be interested in [1], [3], [4], [5], [6] and [8], for example. This section of the book will introduce the reader to three simple programming structures: defining functions, for and while loops and if, elif, else constructs. To open a **New File**, simply click on **File** and **New File** on your computer, see Figure 1.3. Before listing simple examples, the following hints for programming are important to note:

- Indentation: The indentation level in Python code is significant.

- Common typing errors: Include all operators, make sure parentheses match up in correct pairs, check syntax using the **help** command and note that Python is case sensitive.

- Continuation lines: Use a backslash to split code across multiple lines.

- Preallocate arrays: Use the zeros command to save on memory.

- Read the warning messages supplied by Python.

- Check that you are using the correct libraries (modules).

- Keep the program as simple as possible.

- Use the World Wide Web when seeking help.

We will be constructing examples using the following programming constructs:

def	**for**	**while**	**if, elif, else**
def *name(parameters)*:	for *item* in *object*:	while *expression*:	if *expression*:
statement(s)	*statement(s)*	*statement(s)*	*body of if*
			elif *expression*:
			body of elif
			else:
			body of else

Note that the indentation levels in Python are vital to the successful running of the programs. When you hit the RETURN key after the colon, Python will automatically indent for you.

1.2.1 Tutorial Three: Defining Functions (30 Minutes)

The reader can think of this as adding buttons to the powerful Python calculator discussed in the previous sections. Typically, one defines the function, runs the module and then calls the new function in the IDLE Shell or in another program.

Example 1.2.1. Defining the sqr function.

To define a function in Python, one uses the **def** command. Figure 1.4 shows the first function that has been saved as sqr.py on the computer. Note that the comments are colored in red text, and the green text between the pair of triple quotes can be used to give more information. Just like the hash symbol, Python ignores text between the pairs of triple quotes. The reader types **def sqr(x)**, followed by a colon. On hitting the RETURN key, Python automatically indents to the next line, where the reader types **return x * x**. Note that the indentation levels in Python are vitally important and more examples will follow.

Figure 1.5 shows the output after running the module in the Editor Window and typing >>> `sqr(-9)` in the IDLE Shell Window.

Example 1.2.2. Defining the logistic map function.

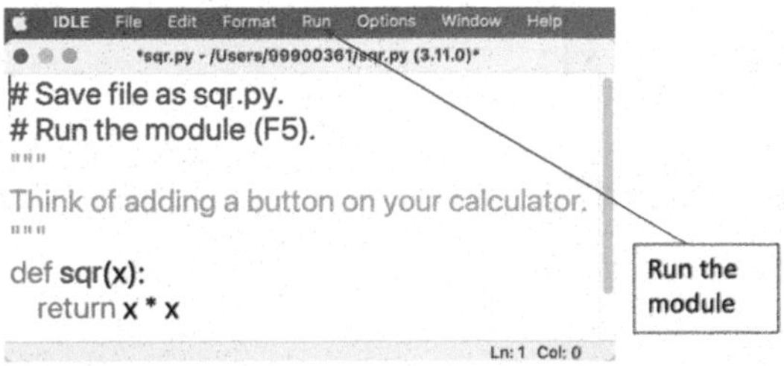

Figure 1.4 The function **sqr** that returns the square of a number. To call the function, one needs to Run the module (or type F5).

```
● ● ●    sqr.py – Editor Window
# The square function – save file as sqr.py.
# Run the Module (or type F5).
"""
This is our first Python program.
Think of adding a square button onto your
Python calculator.
"""
def sqr(x):
    return x * x
```
```
● ● ●      Python Shell
>>> sqr(-9)
81
```

Figure 1.5 Run the module before calling the function in the IDLE shell.

Figure 1.6 shows the program for defining the logistic map function, which is a function of two variables, μ and x. Note that the output is slightly incorrect as $4 \times 0.2 \times (1 - 0.2) = 0.64$. The error is due to floating point arithmetic. In computing, floating-point numbers are represented in computer hardware as base 2 binary fractions. The fraction one-eighth can be expressed as an exact binary decimal. In binary, for example, $0.001 = \frac{0}{2} + \frac{0}{4} + \frac{1}{8}$. However, the fraction $\frac{1}{10}$ cannot be expressed exactly as a base 2 decimal. In the case of $\frac{1}{10}$, for example, Python approximates using the binary fraction $\frac{1}{10} \approx 3602879701896397/2^{55}$. Later in the book, readers will discover that these small rounding errors can greatly affect the outcome of numerical computation, leading to incorrect results. One way to overcome these errors is to use symbolic computation using sympy (see Chapter 2).

Example 1.2.3. A function to convert degrees Fahrenheit into Kelvin.

Figure 1.7 lists the Python code. Define a function called F2K(), where the empty brackets indicate that this is a function with an empty place-holder, and remember to put a colon at the end of the line. Python indents, and we assign F to be of class float and then IDLE will prompt for input from the user using the input command.

```
● ● ●    f_mu.py – Editor Window
# The logistic function – save file as f_mu.py.
# Run the Module (or type F5).
def f_mu(mu , x):
    return mu * x * (1 – x)
```
```
● ● ●      Python Shell
>>> f_mu(4 , 0.2)
0.6400000000000001
```

Figure 1.6 Run the module before calling the function in the IDLE Python shell. There is a slight error due to floating point arithmetic.

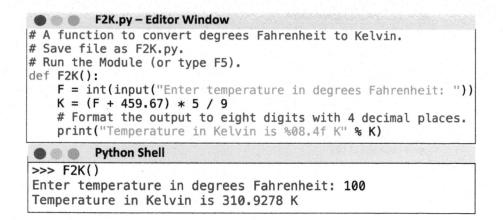

```
● ● ●   F2K.py – Editor Window
# A function to convert degrees Fahrenheit to Kelvin.
# Save file as F2K.py.
# Run the Module (or type F5).
def F2K():
    F = int(input("Enter temperature in degrees Fahrenheit: "))
    K = (F + 459.67) * 5 / 9
    # Format the output to eight digits with 4 decimal places.
    print("Temperature in Kelvin is %08.4f K" % K)
```

```
● ● ●   Python Shell
>>> F2K()
Enter temperature in degrees Fahrenheit: 100
Temperature in Kelvin is 310.9278 K
```

Figure 1.7 Run the module before calling the function. The input and output are shown in the IDLE Python Shell.

The temperature in Kelvin is computed in the next line, and finally, the output is printed using the format %08.4f, which gives eight place holders and four decimal places.

1.2.2 Tutorial Four: For and While Loops (20 Minutes)

Example 1.2.4. Print the first n terms of the Fibonacci sequence using a **for** loop.

Figure 1.8 lists the Python code and output in the IDLE Shell. Note that in Python one can assign multiple values in one line: a, b = 0, 1, for instance. This program defines a function called **Fibonacci**.

Example 1.2.5. Write a program to sum the first N natural numbers using a **while** loop.

Figure 1.9 lists the Python code and output in the IDLE Python Shell. Note that one may increment a counter using i += 1, which is equivalent to i = i + 1.

1.2.3 Tutorial Five: If, elif, else constructs (10 Minutes)

Example 1.2.6. Write a program to grade student results using **if, elif, else**.

Figure 1.10 lists the Python code and output in the IDLE Shell. This program gives a function called **Grade**.

1.3   THE TURTLE MODULE AND FRACTALS

There are a few independently published books on the turtle module, see for example, [2] and [7]. The three programming constructs introduced in the previous section will now be used to plot simple fractals using the turtle module. Drawing pictures using

```
● ● ●     Fibonacci.py – Editor
# A function to list the n terms of the Fibonacci sequence.
# Save file as Fibonacci.py.
def Fibonacci(n):
    a , b = 0 , 1
    print(a) , print(b) , print(a+b)
    for i in range(n-3):
        a , b = b , a + b
        print(a + b)
```

```
● ● ●     Python Shell
>>> Fibonacci(10)
0
1
1
2
3
5
8
13
21
34
```

Figure 1.8 Run the module before calling the function in the IDLE Python Shell. A program to list the first n terms of the Fibonacci sequence.

iterative processes will give the reader a clearer understanding of programming. More examples can be found under the **Help** tab of the IDLE window (see Figure 1.2), simply click on **Turtle Demo**.

Example 1.3.1. The Cantor fractal is obtained by taking a straight line segment at stage zero and removing the middle third segment to obtain stage one. At each subsequent stage, middle third segments are removed, thus at stage one there are two segments, at stage two there are four segments, and so on. Plot the Cantor set up to stage 4 using the turtle module.

```
● ● ●     sumN.py – Editor Window
# A program that sums the natural numbers to N.
# Save file as sumN.py.
def sumN(n):
    sum , i = 0 , 1
    while i <= n:
        sum += i    # sum = sum + i
        i += 1      # Update counter.

    print("The sum is", sum)
```

```
● ● ●     Python Shell
>>> sumN(100)
The sum is 5050
```

Figure 1.9 Run the module before calling the function in the IDLE shell. A program to sum the first N natural numbers.

```
● ● ●      Grade.py – Editor Window
# A program that grades scores.
# Save file as Grade.py.
# Run the Module (or type F5).
def Grade(score):
    if score >= 70:
        letter = "A"
    elif score >= 60:
        letter = "B"
    elif score >= 50:
        letter = "C"
    elif score >= 40:
        letter = "D"
    else:
        letter = "F"
    return letter
```

```
● ● ●      Python Shell
>>> Grade(85)
'A'
```

Figure 1.10 Run the module before calling the function in the IDLE shell. A program to grade student results.

Figure 1.11 lists the Python code and fractal plotted in Figure 1.12.

To understand how the program works, it is recommended to go through the code slowly, line by line, and plot the fractal by hand, working out the x and y values as you progress. Thus, plot the fractal up to stage 2, and change the speed of the turtle to **speed(1)**, the slowest speed.

```
● ● ●      cantor.py – Editor Window
# Cantor fractal.
# Run Module and type >>> cantor(-400, 200 , 4) in Python Shell.
from turtle import *            # Load functions from turtle module.
speed(0)                        # Fastest speeed of turtle.
def cantor(x, y, length):
    if length >= 5:             # Stop when length < 5.
        penup()
        pensize(3)
        pencolor("blue")
        setpos(x , y)
        pendown()
        fd(length)
        y -= 60                 # Height difference.
        cantor(x , y, length / 3)
        cantor(x + 2 * length / 3, y, length / 3)
        penup()
        setpos(x , y + 60)
```

```
● ● ●      Python Shell
>>> cantor(-400 , 200 , 729)
```

Figure 1.11 Run the module before calling the function in the IDLE shell. A program to plot the Cantor set.

Figure 1.12 The Cantor set plotted in Python.

Example 1.3.2. The Koch curve fractal is plotted using the program listed in Figure 1.13. To see a motif of the fractal set level to one. This motif is applied to smaller and smaller segments to obtain the fractal. Thus at each stage, one segment is replaced by four segments, each one-third the length of the previous segment.

Figure 1.13 lists the Python code and fractal plotted in Figure 1.14. Again, to gain an understanding of how the program works, the reader is encouraged to go through the program line-by-line and plot the fractal by hand. It helps if you first set speed(1), so that the turtle moves at its slowest speed.

Example 1.3.3. Use the turtle module to plot a colorful bifurcating fractal tree.

● ● ●   **KochCurve.py – Editor**

```
# Koch curve fractal.
# Run Module and type >>> KochCurve(300,4) in Python Shell.
from turtle import *
speed(0)                                    # Fastest speed.
def KochCurve(length,level):
    if level==0:
        fd(length)
        return
    KochCurve(length/3,level-1)
    lt(60)                                  # Left turn 60 degrees.
    KochCurve(length/3,level-1)
    rt(120)                                 # Right turn 120 degrees.
    KochCurve(length/3,level-1)
    lt(60)
    KochCurve(length/3,level-1)
```

● ● ●   **Python Shell**

```
>>> KochCurve(300 , 4)
```

Figure 1.13 Run the module before calling the function in the IDLE shell. A program to plot the Koch curve fractal. The upper figure shows the motif obtained by setting stage is equal to one.

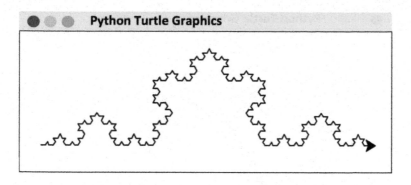

Figure 1.14 The Koch curve plotted with Python. To get the motif, type KochCurve(300 , 1) in the Python shell.

Figure 1.15 lists the Python code, and Figure 1.16 shows the fractal plotted in the Python Turtle Graphics window.

Example 1.3.4. The Sierpinski triangle fractal is obtained by taking an equilateral triangle and removing the middle inverted triangle at each stage. Use the turtle module to plot a Sierpinski triangle to stage 4.

```
● ● ●    FractalTreeColor.py – Editor Window
# A program to plot a color fractal tree.
# Run Module and type >>> FractalTreeColor(200,10) in Python Shell.
from turtle import *
setheading(90)              # The turtle points straight up.
penup()                     # Lift the pen.
setpos(0,-250)              # Set initial point.
pendown()                   # Pen down.
def FractalTreeColor(length, level):
    pensize(length/10)      # Thickness of lines.
    if length < 20:
        pencolor("green")
    else:
        pencolor("brown")
    speed(0)
    if level > 0:
        fd(length)          # Forward.
        rt(30)              # Right turn 30 degrees.
        FractalTreeColor(length*0.7, level-1)
        lt(90)              # Left turn 90 degrees.
        FractalTreeColor(length*0.5, level-1)
        rt(60)              # Right turn 60 degrees.
        penup()
        bk(length)          # Backward.
        pendown()
```

```
● ● ●    Python Shell
>>> FractalTreeColor(200 , 10)
```

Figure 1.15 Run the module before calling the function in the IDLE shell. A program to plot a bifurcating tree.

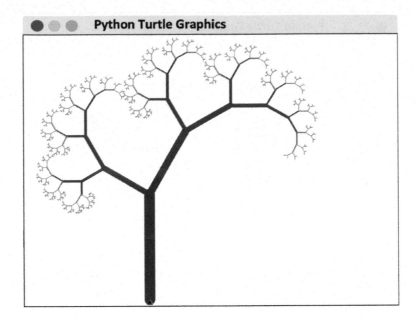

Figure 1.16 A color bifurcating tree.

Figure 1.17 lists the Python code, and the fractal is plotted in the Python Turtle Graphics window in Figure 1.18.

```
● ● ●    Sierpinski.py – Editor Window
# Sierpinski triangle - save file as Sierpinski.py.
# Run the Module (or type F5).
# In the Python shell, type >>> Sierpinski(400 , 5).
from turtle import *
def Sierpinski(length, level):
    speed(0)        # Fastest speed.
    if level==0:
        return
    begin_fill()    # Fill shape.
    color("red")

    for i in range(3):
        Sierpinski(length/2,level-1)
        fd(length)
        lt(120)     # Left turn 120 degrees.
    end_fill()
```

```
● ● ●    Python Shell
>>> Sierpinski(400 , 5)
```

Figure 1.17 Run the module before calling the function in the IDLE shell. A program to plot a red colored Sierpinski triangle. Note that to get stage 4, one needs to type Sierpinski(200, 5), in the Console Window for this program.

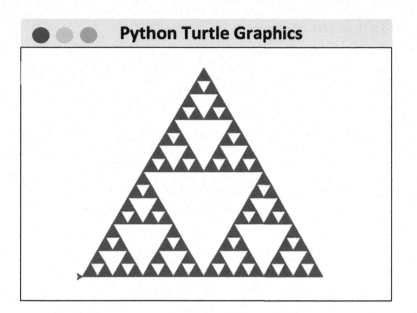

Figure 1.18 The Sierpinski triangle fractal.

Fractals will be discussed in far more detail in a later chapter of the book.

EXERCISES

1.1 Using Python as a powerful calculator with the IDLE Shell, compute:

(a) $2 \times (3 - 4 \times (5 - 8))$;

(b) $e^{\sin(30^o)}$;

(c) $\mathrm{floor}(\pi) - \mathrm{ceil}(e)$;

(d) $\frac{3602879701896397}{2^{55}}$, as a floating point number;

(e) $\frac{4}{5} - \frac{1}{7} \times \frac{2}{3}$, as a fraction.

1.2 Lists:

(a) Construct a list of even integers from 4 to 400.

(b) Construct a list of the form $[-9, -5, -1, \ldots, 195, 199]$.

(c) Given the list, A = [[1,2,3],[4,5,6],[7,8,9]], how would you access the third element in the second list?

(d) Given the list, a = [10,3,4,7,8,2,5,3,4,12], determine the mean of this list using the functions **sum** and **len**.

1.3 Simple Python programming:

(a) Write a function for converting degrees Fahrenheit in to degrees Centigrade.

(b) Write a Python program that sums the subset of prime numbers up to some natural number, n, say.

(c) Write an interactive Python program to play a "guess the number" game. The computer should think of a random integer between 1 and 20, and the user (player) has to try to guess the number within six attempts. The program should let the player know if the guess is too high or too low. Readers will need the randint function from the random module, examples and syntax can be found on the internet.

(d) Consider Pythagorean triples, positive integers a, b, c, such that $a^2 + b^2 = c^2$. Suppose that c is defined by $c = b + n$, where n is also an integer. Write a Python program that will find all such triples for a given value of n, where both a and b are less than or equal to a maximum value, m, say. For the case $n = 1$, find all triples with $1 \leq a \leq 100$ and $1 \leq b \leq 100$. For the case $n = 3$, find all triples with $1 \leq a \leq 200$ and $1 \leq b \leq 200$.

(e) Write a program to output a list of words that are longer in length than a given number, n, say, from an input sentence. For example, the word apple is of length five.

1.4 Plotting fractals with the turtle module:

(a) Edit the Cantor set program to plot a variation of the Cantor fractal, where two middle segments, each of length one-fifth are removed at each stage. Thus, at stage one, there will be three segments each of length one-fifth and at stage two there will be nine segments each of length $\frac{1}{25}$.

(b) Edit the Koch curve Python program to plot a Koch square curve, where each segment is replaced with 5 segments, each one-third the length of the previous segment.

(c) Edit the program for plotting a bifurcating tree to plot a trifurcating tree, where at each stage of the construction, three branches are formed.

(d) Edit the sierpinski.py Python program to construct a Sierpinski square fractal, where the central square is removed at each stage and the length scales decrease by one third.

Solutions to the Exercises may be found here:

https://drstephenlynch.github.io/webpages/Solutions_Section_1.html.

FURTHER READING

[1] Ceder, N. (2018). *The Quick Python Book*. Manning Publications, New York.

[2] Farrell, P.A. (2015). *Hacking Math Class with Python: Exploring Math Through Computer Programming*. CreateSpace Independent Publishing Platform.

[3] Ledger, L.J. (2022). *Python Programming for Beginners*. Independently published.

[4] Mathes, E. (2019). *Python Crash Course: A Hands-On, Project-Based Introduction to Programming 2nd Edition*. No Starch Press.

[5] Ozoemena, S. (2021). *Python for Beginners: Learn Python Programming With No Coding Experience in 7 Days*. Independently Published.

[6] Shovic, J.C. and Simpson, A. (2021). *Python All-in-One For Dummies (For Dummies (Computer/Tech)) 2nd Edition*. For Dummies.

[7] Vashishtha, T. (2021). *My Python Turtles*. Independently Published.

[8] Venkitachalam, M. (2023). *Python Playground, 2nd Edition*. No Starch Press.

## CHAPTER 2

Anaconda, Spyder and the Libraries NumPy, Matplotlib and SymPy

Anaconda Individual Edition is currently the world's most popular Python distribution platform. To download Anaconda for Windows, Mac or Linux, use this URL:

`https://www.anaconda.com/products/individual`.

Anaconda Navigator is a Graphical User Interface (GUI) from which users can launch applications. Figure 2.1 shows the Anaconda GUI displaying the applications JupyterLab, jupyter Notebook, IP[y] Qt Console, Spyder, Glueviz, Orange 3 and RStudio. In this book, we will be using Spyder and jupyter notebooks only. JupyterLab is the next-generation user interface and will eventually replace jupyter notebooks; the Qt Console is a very lightweight application that appears like a terminal; Glueviz is an interface to explore relationships within and between related datasets; Orange 3 is a data visualization and data analysis toolbox and RStudio is an IDE for the statistics package R.

To launch Spyder, simply click on the Launch button as indicated in Figure 2.1. Figure 2.2 shows the Integrated Development Environment (IDE) for Spyder. Spyder is an acronym for "Scientific PYthon Development EnviRonment." The Editor Window (on the left) is used to write code (programs) and save to file. The IPython Console Window (bottom right) is where the user can use Python as a powerful graphing calculator. The command prompt "In [1]:" indicates that the console is waiting for you to type commands. The Variable explorer/Help/Plots/Files window (top right) can display detailed help on variables or files and is useful when debugging, it is also where your plots will be displayed. At any time, the user can undock any of the windows by clicking on the three horizontal bars in the top right-hand corners of the windows. You can also set **Preferences** in Spyder, for example, font size and style, and you can elect to work in dark mode or light mode. I have used light mode in this book.

DOI: 10.1201/9781003285816-2

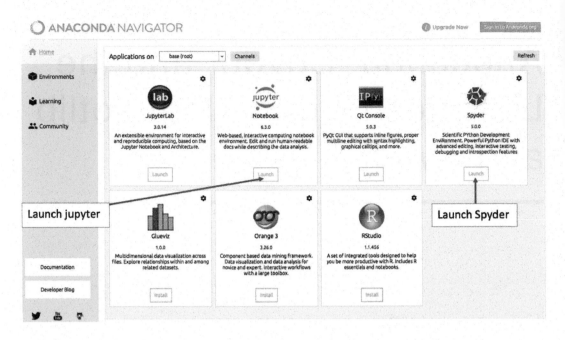

Figure 2.1 The Anaconda user interface. We will only use Spyder and jupyter Notebooks in this text.

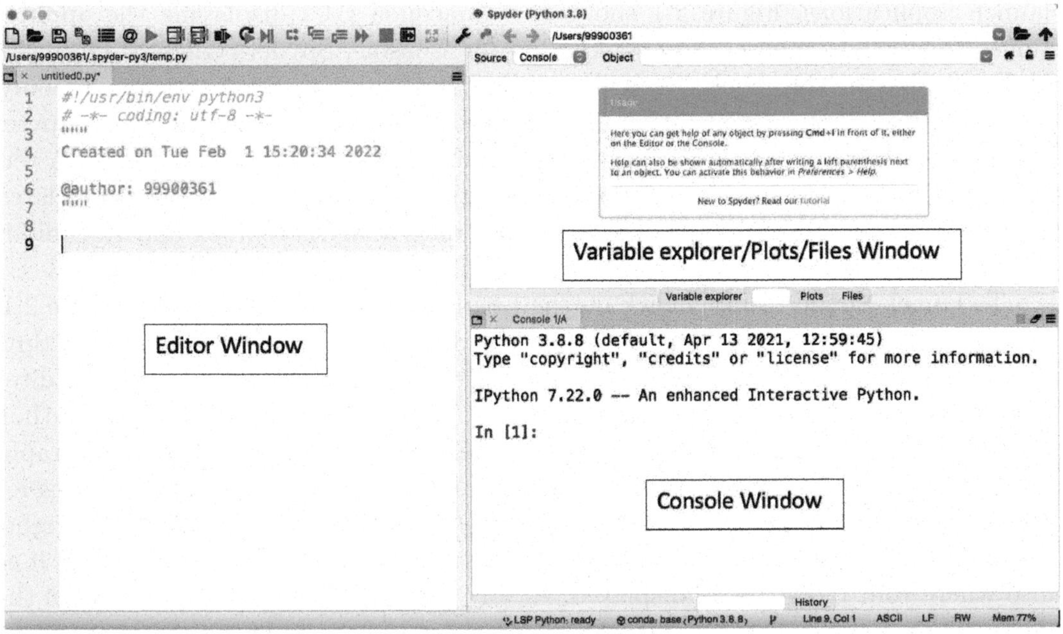

Figure 2.2 The Spyder Integrated Development Environment (IDE) in light mode. Those of you familiar with MATLAB will see similarities here.

Aims and Objectives:

- To use Anaconda to program Python.

- To introduce the numerical python package NumPy (or numpy).

- To introduce the MatPlotLib (or matplotlib) plotting package.

- To introduce symbolic computation using SymPy (or sympy).

At the end of the chapter, the reader should be able to

- install Anaconda and launch Spyder and jupyter notebooks;

- manipulate lists, arrays, vectors, matrices and tensors;

- plot simple curves and annotate;

- carry out simple symbolic computation.

2.1 A TUTORIAL INTRODUCTION TO NUMPY

NumPy is a library that allows Python to compute with lists, arrays, vectors, matrices and tensors. Recall that lists were introduced in Chapter 1. For more detail, the reader is directed to [1] and to the URL:

`https://docs.scipy.org/doc/numpy/reference/`.

2.1.1 Tutorial One: An Introduction to NumPy and Arrays (30 Minutes)

The reader should copy the commands after the prompt In[n]: in the Console Window (see Figure 2.2). There is no need to copy the comments, they are there to help you. The reader is encouraged to make up their own examples in order to fully understand the command lines below. Important notes are listed below the list of command lines.

Python Command Lines   **Comments**

```
In[1]: import numpy as np          # Import numpy into the np namespace.
In[2]: A = np.arange(5)            # An array, A=[0 1 2 3 4].
In[3]: print(A)                    # Print the array. Note, no commas.
In[4]: A.tolist()                  # Convert an array to a list.
In[5]: print(Out[4])               # You can use previous Output.
In[6]: u = np.array([1 , 2 , 3])   # A 3D vector and rank one tensor.
In[7]: v = np.array([4 , 5 , 6])   # A 3D vector.
In[8]: np.dot(u , v)               # The dot product of two vectors.
In[9]: np.cross(u , v)             # The cross product of two vectors.
```

```
In[10]: B = np.array([[1,1],[0,1]])      # A 2D array, rank 2 tensor.
In[11]: C = np.array([[2,0],[3,4]])      # A 2D array.
In[12]: B * C                            # Elementwise product.
In[13]: np.dot(B , C)                     # Matrix product.
In[14]: D=np.arange(9).reshape(3,3)      # A 2D array.
In[15]: D.sum(axis = 0)                   # Sum columns, vectorized computation.
In[16]: D.max(axis = 0)                   # The maximum of each column.
In[17]: D.min(axis = 1)                   # The minimum of each row.
In[18]: D.cumsum(axis = 1)                # Cumulative sum of each row.
In[19]: type(D)                           # numpy.ndarray.
In[20]: D.ndim                            # The tensor rank of D is 2.
In[21]: E = np.zeros((3 , 3))             # Creates a 3 × 3 array of zeros.
In[22]: np.ones((3 , 3))                  # Creates a 3 × 3 array of ones.
In[23]: np.full((3 , 3) , 7)              # Creates a 3 × 3 array of 7s.
In[24]: I = np.eye(3)                     # Creates a 3 × 3 identity matrix.
In[25]: M=np.arange(12).reshape(3,4)     # A 3 × 4 array.
In[26]: M[1 , 3]                           # The element in row 2 column 4.
In[27]: M[: , 1]                           # Column 2, array([1,5,9]).
In[28]: N = M[: 2 , 1 : 3]                 # Slice to get subarray.
In[29]: print(N)                           # First 2 rows, and cols 2 and 3.
In[30]: quit                               # Restart the kernel.
```

Important notes:

- The numpy library (or module) is called using the alias np. The simple reason for this is that all Python programmers use this aliasing.

- Python uses zero-based indexing.

- The elements in lists are separated by commas in Python.

- The elements in arrays are separated by gaps in Python.

- Make up your own examples to get a better understanding of how slicing works. Slicing is also used when vectorizing a program (see Chapter 12).

- The command **ndim** is used to find the tensor rank of an array. Tensors will be explained in more detail later in the book (see TensorFlow for machine learning and deep learning).

- Just as with IDLE, the user can use the up and down arrow keys to return to previously typed command lines.

Vectorized Operations in Numpy. There is a single data type in numpy arrays and this allows for vectorization. As far as the reader is concerned, for loops, which can use a lot of memory in programs, can be replaced with vectorized forms. As a simple example, consider the list of numbers, $a = [1, 2, 3, \ldots, 100]$, and suppose we wish to compute $\sum_{i=1}^{100} (a_i^2 + 1)$. Then, using for loops, the code would be:

```
sum = 0
a= [i for i in range(101)]
for j in range(101):
    sum += a[j]**2 + 1
print(sum)
```

The equivalent, vectorized form is:

```
import numpy as np
a=np.arange(101)
print("The sum is" , np.sum(a**2+1))
```

More examples of vectorized programs are listed in Chapter 12. For huge data sets, vectorization can speed up programs by several orders of magnitude.

2.2 A TUTORIAL INTRODUCTION TO MATPLOTLIB

Matplotlib is an acronym for MATrix PLOTting LIBrary and it is a comprehensive library for creating animated, static, and more recently, interactive visualizations in Python. For more information the reader is directed to the URL:

```
https://matplotlib.orgl.
```

2.2.1 Tutorial Two: Simple Plots using the Spyder Editor Window (30 minutes)

Simply copy the commands into the Editor Window and save the file as you would in any Windows document. To run the file, click on the green play button in the tab bar (or type F5). The plot will appear in the Plots Window in the upper right of the IDE.

Important notes:

- The matplotlib.pyplot library (or module) is always called using the alias plt.

- The function np.linspace creates a vector of domain values.

- LaTeX is used to get the $y = x^2$ title. LaTeX will be discussed in the next chapter.

- You can change the resolution (dpi=dots (pixels) per inch) of a figure using **dpi** in the **savefig** function. See Example 2.2.1.1.

- Always use **plt.show()** to show your plot.

Example 2.2.1.1. Plot the curve $y = x^2$. See plot1.py.

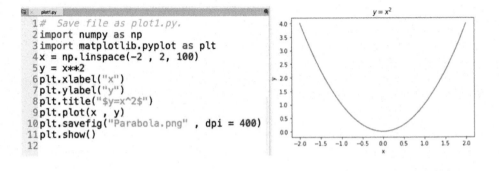

Example 2.2.1.2. Plot the parametric curve given by $y(t) = 0.7 \sin(t + 1) \sin(5t)$ and $x(t) = 0.7 \cos(t + 1) \sin(5t)$. See plot2.py.

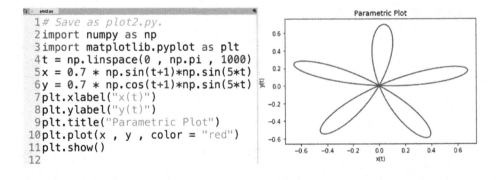

Example 2.2.1.3. Plot the curves $c = 1 + \cos(2\pi t)$ and $s = 1 + \sin(2\pi t)$ on the same grid graph. Note that **savefig** is used to save a plot on your computer. The file will appear in the same folder as **plot3.py**. See plot3.py.

Subplot. The next example illustrates how one may obtain subplots. The syntax for the **subplot** command is **subplot(m , n , i)**, where m is the number of rows, n is the number of columns, and i is the number of the figure. In Example 2.2.1.4, there are two rows and one column in the subplot.

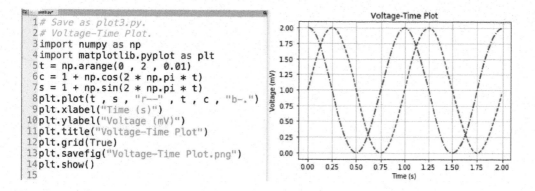

```
1 # Save as plot3.py.
2 # Voltage-Time Plot.
3 import numpy as np
4 import matplotlib.pyplot as plt
5 t = np.arange(0 , 2 , 0.01)
6 c = 1 + np.cos(2 * np.pi * t)
7 s = 1 + np.sin(2 * np.pi * t)
8 plt.plot(t , s , "r--" , t , c , "b-.")
9 plt.xlabel("Time (s)")
10 plt.ylabel("Voltage (mV)")
11 plt.title("Voltage-Time Plot")
12 plt.grid(True)
13 plt.savefig("Voltage-Time Plot.png")
14 plt.show()
15
```

Example 2.2.1.4. Use the **subplot** function to plot the solutions to a damped and undamped pendulum.

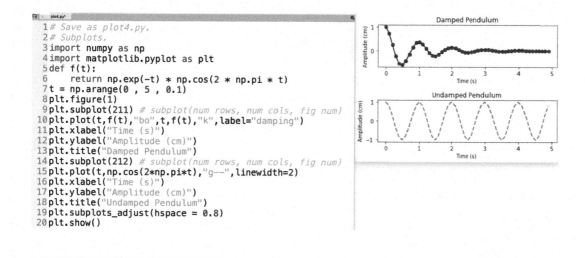

```
1 # Save as plot4.py.
2 # Subplots.
3 import numpy as np
4 import matplotlib.pyplot as plt
5 def f(t):
6     return np.exp(-t) * np.cos(2 * np.pi * t)
7 t = np.arange(0 , 5 , 0.1)
8 plt.figure(1)
9 plt.subplot(211) # subplot(num rows, num cols, fig num)
10 plt.plot(t,f(t),"bo",t,f(t),"k",label="damping")
11 plt.xlabel("Time (s)")
12 plt.ylabel("Amplitude (cm)")
13 plt.title("Damped Pendulum")
14 plt.subplot(212) # subplot(num rows, num cols, fig num)
15 plt.plot(t,np.cos(2*np.pi*t),"g--",linewidth=2)
16 plt.xlabel("Time (s)")
17 plt.ylabel("Amplitude (cm)")
18 plt.title("Undamped Pendulum")
19 plt.subplots_adjust(hspace = 0.8)
20 plt.show()
```

Example 2.2.1.5. Load data from an Excel spreadsheet. Plot a graph of voltage against current, and determine a polynomial (of degree 20) of best fit to the data. See plot5.py.

The Excel spreadsheet **Kam_Data_nA.xlsx** must be in the same folder as the Python file **plot5.py**. Pandas is a fast, powerful, flexible and easy to use open source data analysis and manipulation tool for Python. The xs values are taken from the column headed **Volt**, and the ys values are taken from the column headed **nAJ**, from the Excel spreadsheet. The **polyfit** function from the numpy library is used to find a polynomial (of degree 20 in this case). The polynomial of best fit will be printed when the program is run. The reader is encouraged to plot the polynomial and data points on the same graph to see how well the polynomial fits the data. The data file can be downloaded through the books' GitHub repository.

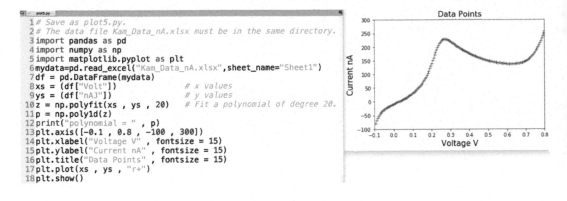

```
1 # Save as plot5.py.
2 # The data file Kam_Data_nA.xlsx must be in the same directory.
3 import pandas as pd
4 import numpy as np
5 import matplotlib.pyplot as plt
6 mydata=pd.read_excel("Kam_Data_nA.xlsx",sheet_name="Sheet1")
7 df = pd.DataFrame(mydata)
8 xs = (df["Volt"])                # x values
9 ys = (df["nAJ"])                 # y values
10 z = np.polyfit(xs , ys , 20)    # Fit a polynomial of degree 20.
11 p = np.poly1d(z)
12 print("polynomial = " , p)
13 plt.axis([-0.1 , 0.8 , -100 , 300])
14 plt.xlabel("Voltage V" , fontsize = 15)
15 plt.ylabel("Current nA" , fontsize = 15)
16 plt.title("Data Points" , fontsize = 15)
17 plt.plot(xs , ys , "r+")
18 plt.show()
```

2.3   A TUTORIAL INTRODUCTION TO SYMPY

SymPy is a computer algebra system and a Python library for symbolic mathematics written entirely in Python. For more information, see [3] and the sympy help pages at:

https://docs.sympy.org/latest/index.html.

2.3.1   Tutorial Three: An Introduction to SymPy (30 Minutes)

The reader should copy the commands after the prompt In[n]: in the Console Window (see Figure 2.2). There is no need to copy the comments, they are there to help you.

Python Command Lines	Comments
In[1]: from sympy import *	# Import all functions.
In[2]: n , x , y = symbols("n x y")	# Declare n, x, y symbolic.
In[3]: factor(x**2 - y**2)	# Factorize.
In[4]: solve(x**2 - 4 * x - 3 , x)	# Solve an algebraic equation.
In[5]: apart(1 / ((x+2) * (x+1)))	# Partial fractions.
In[6]: trigsimp(cos(x) - cos(x)**3)	# Simplify trig expressions.
In[7]: limit(x / sin(x) ,x , 0)	# Limits.
In[8]: diff(x**2 - 7 * x + 8 , x)	# Differentiation.
In[9]: diff(5 * x**7 , x , 3)	# The third derivative.
In[10]: diff(3*x**4*y**7,x,2,y,1)	# Partial differentiation.
In[11]: (exp(x)*cos(x)).series(x,0,10)	# Taylor series expansion.
In[12]: integrate(x**4 , x)	# Indefinite integration.
In[13]: integrate(x**4 , (x,1,3))	# Definite integration.
In[14]: integrate(1 / x**2, (x,1,oo))	# Improper integration.
In[15]: summation(1/n**2,(n,1,oo))	# Infinite sum.

```
In[16]: solve([x-y,x+2*y-5],[x,y])      # Linear simultaneous equations.

In[17]: solve([x-y,x**2+y**2-1],[x,y])  # Nonlinear simultaneous equations.

In[18]: N(pi , 500)                      # π to 500 decimal places.

In[19]: A = Matrix([[1,-1],[2,3]])       # A 2 × 2 matrix.

In[20]: B = Matrix([[0,2],[3,3]])        # A matrix.

In[21]: 2 * A + 3 * B                    # Matrix algebra.

In[22]: A * B                            # Matrix multiplication.

In[23]: A.row(0)                         # Access row 1 of A.

In[24]: A.col(1)                         # Access column 2 of A.

In[25]: A[0 , 1]                         # The element in row 1, column 2.

In[26]: A.T                              # The transpose matrix.

In[27]: A.inv()                          # The inverse matrix, if it exists.

In[28]: A.det()                          # The determinant.

In[29]: zeros(5,5)                       # A 5 × 5 matrix of zeros.

In[30]: ones(1 , 5)                      # A 1 × 5 matrix of ones.

In[31]: eye(10)                          # The 10 × 10 identity matrix.

In[32]: M = Matrix([[1,-1],[2,4]])       # A 2x2 matrix.

In[33]: M.eigenvals()                    # Eigenvalues of M are {2:1,3:1}.

In[34]: z1 , z2 = 3 + 1j , 5 - 4j        # Complex numbers. See notes below.

In[35]: 2 * z1 + z1 * z2                 # Complex algebra.

In[36]: abs(z1)                          # The modulus.

In[37]: re(z1) , im(z1)                  # Real and imaginary parts.

In[38]: arg(z1)                          # The argument.

In[39]: exp(1j*z1).expand(complex=True)  # Express in the form x+jy.

In[40]: quit                             # Restart the kernel.
```

Important notes:

- When using sympy, the convention is to load all functions.

- For more information on calculus (differentiation and integration) and linear algebra and matrices, the reader is directed to the books at the end of the chapter, see [2] and [4], respectively.

- The eigenvalues of the matrix M are in the form $\{2 : 1, 3 : 1\}$, meaning that eigenvalue 2 has multiplicity one, and eigenvalue 3 has multiplicity one. Eigenvalues can be used to determine the stability of fixed points (for discrete systems) and the stability of critical points (for continuous systems).

- In sympy, $j = \sqrt{-1}$. For an introduction to complex numbers, and pure mathematics in general, the reader is directed to [5].

- Symbolic computation is quickly becoming one of the most important and defining technologies in this era.

EXERCISES

2.1 Use the command $A = $ np.arange(100).reshape(10, 10), to create a 10×10 array.

   (a) Determine the sum of each row of A. What would the non-vectorized program look like?

   (b) Find the maximum of each row of A.

   (c) Find the cumulative sum of the columns of A.

   (d) Determine the element in row 10 and column 5 of A.

   (e) Create a rank 3 tensor, a $2 \times 2 \times 2$ array of your choice. Think of two lots of 2×2 arrays. Tensors will be discussed in more detail later in the book.

2.2 Plot the following functions using matplotlib:

   (a) $y = x^2 - 3x - 18$;

   (b) $y = \cos(2x)$;

   (c) $y = \sin^2(x)$;

   (d) $y = 4x^3 - 3x^4$;

   (e) $y = \cosh(x)$.

2.3 Use sympy to:

   (a) factorize $x^3 - y^3$;

   (b) solve $x^2 - 7x - 30 = 0$;

   (c) split into partial fractions $\frac{3x}{(x-1)(x+2)(x-5)}$;

   (d) simplify $\sin^4(x) - 2\cos^2(x)\sin^2(x) + \cos^4(x)$;

   (e) expand $(y + x - 3)(x^2 - y + 4)$.

2.4 Use sympy to determine:

   (a) $\lim_{n \to \infty} \left(1 + \frac{1}{n}\right)^n$;

   (b) $\frac{d}{dx}\left(3x^4 - 6x^3\right)$;

(c) $\frac{d^3}{dx^3}\left(3x^4 - 6x^3\right)$;

(d) $\int_{x=0}^{1} x^2 - 2x - 3\, dx$;

(e) the first 10 terms of the Taylor series of $e^x \sin(x)$ about $x = 1$.

2.5 Use sympy to:

(a) sum $\sum_{n=1}^{\infty} \frac{1}{n}$;

(b) solve the simultaneous equations, $x - y = 2$ and $x + y = 1$;

(c) solve the simultaneous equations, $x^2 - y = 2$ and $x + y = 1$;

(d) sum the arithmetic series, $\sum_{n=1}^{20} 2 + 3(n - 1)$;

(e) sum the geometric series, $\sum_{n=1}^{20} 2 \times 3^n$.

2.6 Given that:

$$A = \begin{pmatrix} -1 & 2 & 4 \\ 0 & 3 & 2 \\ 1 & 4 & 6 \end{pmatrix}, \quad B = \begin{pmatrix} 1 & -1 & 1 \\ 2 & 0 & -1 \\ 1 & -1 & 1 \end{pmatrix},$$

find:

(a) $3A - 5B$;

(b) $A \times B$;

(c) the inverse of A, if it exists;

(d) the first column of A^8;

(e) the eigenvalues of B.

2.7 Given the complex numbers, $z_1 = 1 - 4j$, $z_2 = 5 + 6j$, where $j = \sqrt{-1}$, determine:

(a) $3z_1 - 5z_2$;

(b) the absolute value, $|z_1 z_2|$;

(c) $z_1 e^{z_2}$, in the form $x + jy$;

(d) $\sin(z_1)$ in the form $x + jy$;

(e) z_2 in polar form.

Solutions to the Exercises may be found here:

https://drstephenlynch.github.io/webpages/Solutions_Section_1.html.

FURTHER READING

[1] Johansson, R. (2018). *Numerical Python: Scientific Computing and Data Science Applications with NumPy, SciPy and Matplotlib.* Apress (2nd Ed.).

[2] Ryan, M. (2016). *Calculus for Dummies, 2nd Ed.* For Dummies.

[3] Sandona, D. (2021). *Symbolic Computation with Python and SymPy.* Independently Published.

[4] Sterling, M.J. (2009). *Linear Algebra for Dummies.* For Dummies.

[5] Warner, S. (2018). *Pure Mathematics for Beginners: A Rigorous Introduction to Logic, Set Theory, Abstract Algebra, Number Theory, Real Analysis, Topology, Complex Analysis, and Linear Algebra.* Amazon.

## CHAPTER 3

Jupyter Notebooks and Google Colab

Jupyter notebooks are web-based interactive platforms for coding in Python which are launched from Anaconda. The notebooks are human-readable documents containing both rich text (Markdown) and executable code. When you open a notebook, an associated kernel is automatically launched enabling you to compute with Python. For a more in-depth introduction to jupyter notebooks, see [3]. Google Colab notebooks are jupyter notebooks hosted by Google. To run a jupyter notebook in Google Colab you must have a Google account. This means that you can execute Python programs without having any software at all on your computer.

Aims and Objectives:

- To use jupyter notebooks to code Python.

- To use Google Colab to perform cloud computing.

- To save notebooks on the computer and through GitHub.

At the end of the chapter, the reader should be able to

- run Python code in jupyter notebooks;

- insert text, html code, figures and LaTeX equations in Markdown cells;

- create animations and interactive plots;

- create html pages for publication on the Web.

3.1 JUPYTER NOTEBOOKS, CELLS, CODE AND MARKDOWN

To load a jupyter notebook, simply click Launch through Anaconda (see Figure 2.1). Figure 3.1 shows the page that opens when you launch jupyter notebook. Click on **New**, and **Python 3** to open a new notebook. Figure 3.2 shows a blank notebook. The boxes with arrows indicate how users save a file, change the file name, run the

DOI: 10.1201/9781003285816-3

◌ Jupyter   Quit   Logout

| Files | Running | Clusters |

Select items to perform actions on them.   Upload | New ▾ | ⟳

☐ 0 ▾ ▮ /	Name ↓	Last Modified	File size
☐ ▢ Applications		10 days ago	
☐ ▢ build		3 months ago	
☐ ▢ Creative Cloud Files		2 months ago	
☐ ▢ Desktop			
☐ ▢ Documents			
☐ ▢ Downloads		a day ago	

Click here to open a new Python 3 ipynb notebook

Figure 3.1   The page that opens after you launch jupyter notebook from Anaconda.

cell (one can also hit **SHIFT + ENTER** in the cell window), interrupt the kernel and toggle between Code and Markdown. One may save the file in a variety of formats including as a jupyter notebook (*filename.ipynb*), as a Python file (*filename.py*) or as html (*filename.html*). If you save the file in html format, then you can publish the file on the Web.

Example 3.1.1. Inserting text and figures.

Figure 3.3 shows the reader how to insert titles and import figures from files on their own computer. In the first cell one types **# Inserting Titles**, by hitting **SHIFT + ENTER** in the cell (or hit the Run button) the title appears in the jupyter notebook. To edit

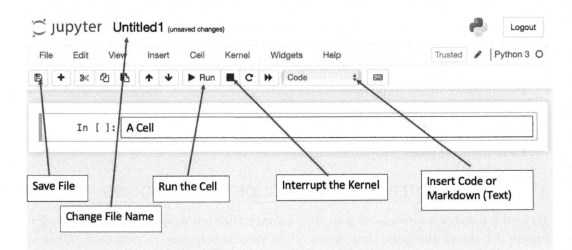

Figure 3.2   A blank, untitled jupyter notebook.

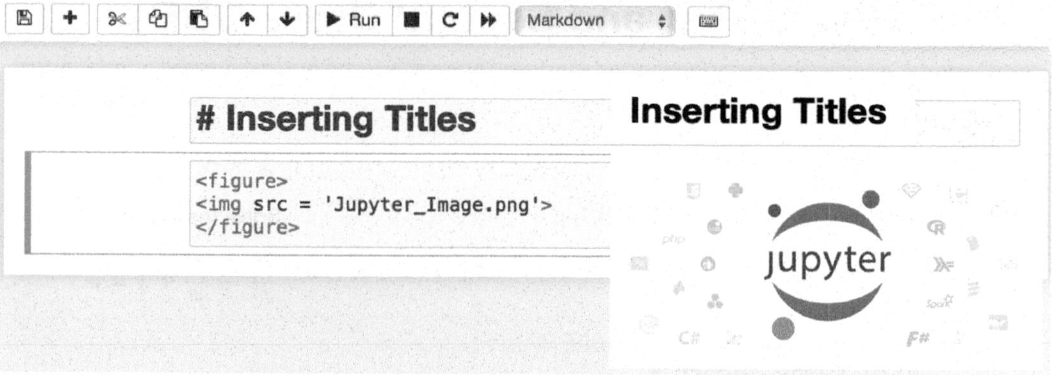

Figure 3.3 Inserting titles and figures using Markdown cells. Ensure that the figure is in the same folder as the jupyter notebook.

the cell one simply double clicks on the cell. In the second cell there is some html code used to display a figure. To import a figure into the notebook from your computer, one uses html code ensuring that the figure file is in the same folder as where the ipynb notebook is saved. As an aside, these cells give the reader an opportunity to learn some basic html for writing web pages.

Example 3.1.2. Inserting equations using LaTeX and solving first and second order ordinary differential equations (ODEs) analytically.

Figure 3.4 shows LaTeX commands in the first cell. The second cell shows the Python code to solve a first order Ordinary Differential Equation (ODE) and the third cell lists the code to solve a second order ODE. The term $\frac{dx}{dt}$ is first order and is read as the rate of change of x with respect to t. Readers will be familiar with speed (or velocity) which is the rate of change of distance with respect to time. The term $\frac{d^2y}{dt^2}$ is second order, and again readers will be familiar with acceleration, which is the rate of change of velocity with respect to time, or the second order rate of change of distance with respect to time. The mathematical methods used to solve differential equations is beyond the remit of this book. We simply use Python to solve them. For readers interested in the methods for solving differential equations, I recommend Schaum's Differential Equations book [1]. Using Python, the solution to the first ODE:

$$\frac{dx}{dt} + x = 1,$$

is

$$x(t) = C1e^{-t} + 1,$$

where $C1$ is a constant. The solution to the second order ODE:

$$\frac{d^2y}{dt^2} + \frac{dy}{dt} + y = e^t,$$

Solving First and Second Order ODEs

Solve the 1st order ODE:

$\frac{dx}{dt}+x=1$

Solve the 2nd order ODE:

$\frac{d^2y}{dt^2}+\frac{dy}{dt}+y=e^t$

Solve the 1st order ODE:

$\frac{dx}{dt} + x = 1$

Solve the 2nd order ODE:

$\frac{d^2y}{dt^2} + \frac{dy}{dt} + y = e^t$

In [1]:
```python
# Solve the 1st order ODE.
from sympy import *
t = symbols('t')
x = symbols('x' , cls = Function)
ODE1 = Eq(x(t).diff(t) , 1 - x(t))    # dx/dt=1-x.
sol1 = dsolve(ODE1 , x(t))
print(sol1)
```

Eq(x(t), C1*exp(-t) + 1)

In [2]:
```python
# Solve the 2nd order ODE.
y = symbols('y' , cls = Function)
ODE2 = Eq(y(t).diff(t,t) + y(t).diff(t) + y(t), exp(t))
sol2 = dsolve(ODE2 , y(t))
print(sol2)
```

Eq(y(t), (C1*sin(sqrt(3)*t/2) + C2*cos(sqrt(3)*t/2))*exp(-t/2)
+ exp(t)/3)

Figure 3.4 Inserting equations using LaTeX and Markdown (top cell). Cells labelled In [1]: and In [2]: are Code cells using Python to solve simple ODEs analytically.

is

$$y(t) = C1 \sin\left(\frac{\sqrt{3}t}{2}\right) + C2 \cos\left(\frac{\sqrt{3}t}{2}\right) e^{-\frac{t}{2}} + \frac{e^t}{3},$$

where $C1$ and $C2$ are constants.

Many more differential equations will be discussed in the second section of the book when modeling real-world systems. Unfortunately, many of these systems do not have nice analytical solutions and numerical methods must be adopted to make any progress.

LaTeX is a document preparation system for high-quality typesetting. I recommend learning LaTeX through Overleaf, an online LaTeX editor:

https://www.overleaf.com.

You have to register to use Overleaf, but there is no installation. There are hundreds of LaTeX templates, for example, assignments, CVs, resumes, journal papers, letters, presentations and theses. Table 3.1 lists some popular LaTeX symbols used

Table 3.1 Table of popular LaTeX symbols for jupyter Notebooks.

Greek Letters	LaTeX	Mathematics	LaTeX
α	\alpha	$\frac{dx}{dt}$	\frac{dx}{dt}
β	\beta	$\dot{x}$	\dot{ x }
γ, Γ	\gamma, \Gamma	$\ddot{x}$	\ddot{x}
δ	\delta	$\sin(x)$	\sin(x)
ϵ	\epsilon	$\cos(x)$	\cos(x)
θ	\theta	$\leq$	\leq
λ, Λ	\lambda, \Lambda	$\geq$	\geq
μ	\mu	x^2	x^2
σ, Σ	\sigma \Sigma	$\in$	\in
τ	\tau	$\pm$	\pm
ϕ	\phi	$\rightarrow$	\rightarrow
ω, Ω	\omega, \Omega	$\int$	\int

in jupyter notebooks. In LaTeX, the commands have to be between dollars. For a more comprehensive list, the reader is directed to:

`https://oeis.org/wiki/List_of_LaTeX_mathematical_symbols`.

3.2 ANIMATIONS AND INTERACTIVE PLOTS

The program below shows an example of how to produce an animation in Python using a jupyter notebook and Figure 3.5 shows the output. This provides a fantastic resource for high school students and teachers. Moving the slider left and right shows how the curve varies as a parameter (frequency in this case) varies. You can show the animation **Once**, **Loop** or **Reflect**. The **Reflect** option is very useful in the realm of dynamical systems as it shows what happens as a parameter increases and then decreases again on a continuous cycle. Many real worlds systems are driven periodically as we live in a periodic solar system and the most common form of solution is periodic. There are exercises at the end of the chapter to give the reader more practice in producing animations. To edit the program, you need only edit the domain and range and the **animate** function.

Example 3.2.1. Use Python to produce an animation of the sine curve, $y = \sin(\omega t)$, as the parameter ω, varies between $0 \leq \omega \leq 5$.

The program is listed below.

```
# Program_3a.ipynb: Animation of a Sine Curve.
# Run notebook in Google Colab.
import numpy as np
import matplotlib.pyplot as plt
from matplotlib import animation, rc
from IPython.display import HTML

# Set up figure.
fig, ax = plt.subplots()
plt.title("Animation of a Sine Wave")
plt.close()
# Set domain and range.
ax.set_xlim(( 0, 2 * np.pi))
ax.set_ylim((- 1.2, 1.2))
# Set line width.
line, = ax.plot([], [], lw=2)
# Initialization function: plot the background of each frame.
def init():
    line.set_data([], [])
    return (line,)
# Animation function. This is called sequentially.
def animate(n):
    x = np.linspace(0, 2 * np.pi, 100)
    y = np.sin(0.05 * n * x)
    line.set_data(x, y)
    return (line,)
# Animate, interval sets the speed of animation.
anim = animation.FuncAnimation(fig, animate, init_func=init,
                               frames=101, interval=100, blit=True)
# Note: below is the part which makes it work on Colab.
rc("animation", html="jshtml")
anim
```

The final program in this chapter, lists a program for producing an interactive plot, which is a relatively new feature in Python.

Example 3.2.2. Use Python to produce an interactive plot of the sine curve, $y = A\sin(Bt + C) + D$, as the parameters vary: $0 \le A \le 5$, $0 \le B \le 2\pi$, $0 \le C \le 2\pi$, and $-3 \le D \le 3$.

The program is listed below.

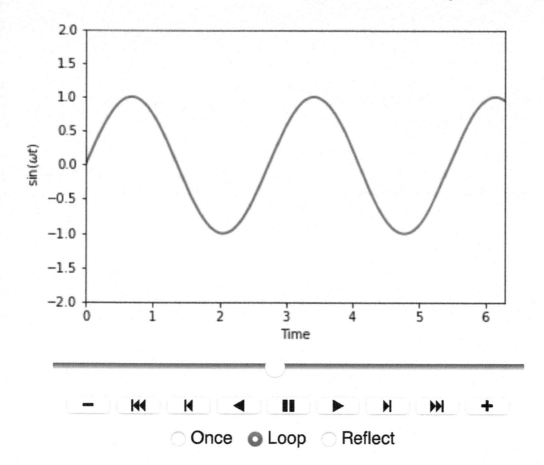

Figure 3.5 An animation of the curve $y = \sin(\omega t)$ as the parameter varies, $0 \leq \omega \leq 5$.

```
# Program_3b.ipynb: Interactive plots with Python.
# Run notebook in Google Colab.
from __future__ import print_function
from ipywidgets import interact, interactive, fixed, interact_manual
import ipywidgets as widgets

%matplotlib inline
from ipywidgets import interactive
import matplotlib.pyplot as plt
import numpy as np
# Interactive plot showing amplitude, frequency, phase and vertical shift.
def f(A, B, C, D):
    plt.figure(2)
    t = np.linspace(-10, 10, num=1000)
    plt.plot(t , A * np.sin(B * (t + C)) + D)
    plt.ylim(-8, 8)
```

```
    plt.show()
interactive_plot = interactive(f, A=(0, 5 , 0.2), B = (0, 2 * np.pi), \
                        C = (0, 2 * np.pi),   D = (-3, 3, 0.5))
output = interactive_plot.children[-1]
output.layout.height = "350px"
interactive_plot
```

Figure 3.6 shows the output. In this case, we plot the curve $y = A\sin(Bt+C)+D$, where A is an amplitude, B is a frequency, C is a phase shift and D is a vertical shift. The parameter A varies from 0 to 5 in steps of 0.2. The parameters B, C vary from 0 to 2π, and D varies from -3 to 3, in steps of 0.5.

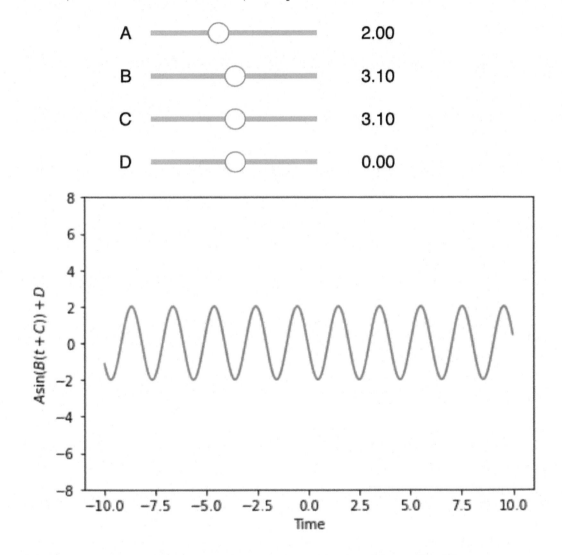

Figure 3.6 An interactive plot of $y = A\sin(Bt + C) + D$.

3.3 GOOGLE COLAB AND GITHUB

Google Colab (Colaboratory) allows the user to write and execute Python code in their browser, with zero configuration required, free access to Graphical Processing Units (GPUs) and Tensor Processing Units (TPUs), and easy file sharing via GitHub. The user must have a Google account to use Colab, you can create a Google account here:

```
https://support.google.com/accounts/.
```

To use Google Colab, click on the link:

```
https://colab.research.google.com/.
```

When you first click on the Google Colab URL above, a live jupyter notebook opens up, as shown in Figure 3.7.

Figure 3.8 shows the web page after signing in to your Google account. Simply click on **New notebook** to open a jupyter notebook

Figure 3.9 shows the new jupyter notebook. There are various options for saving the notebook. Click on **File** to **Save a copy in Drive** or **Save a copy in GitHub**. You can also **File** and **Download** as **.ipynb** or **.py**. To sign up to GitHub, where the world builds software, click here:

```
https://github.com.
```

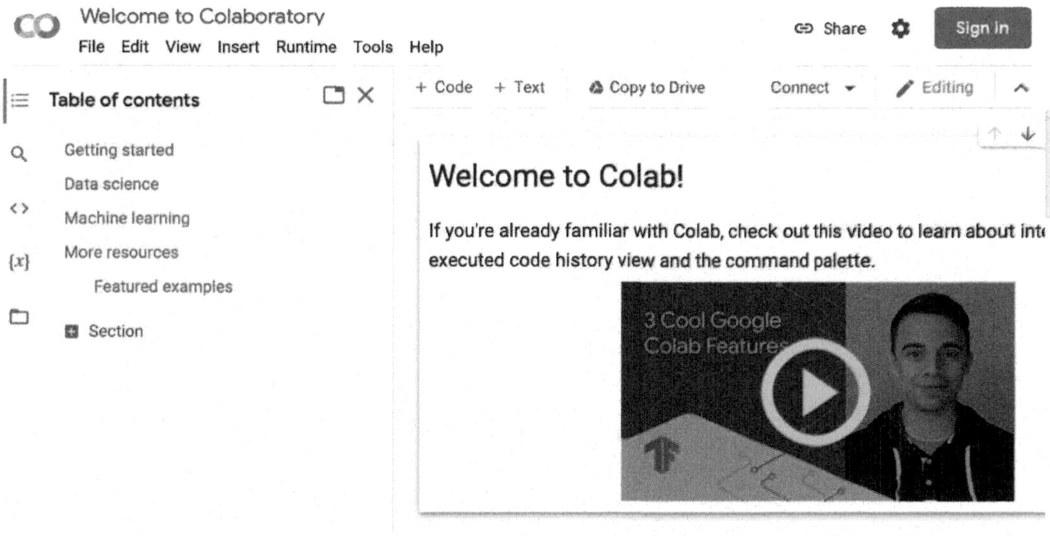

Figure 3.7 The web page that opens up on clicking the Google Colab link. The web page is a live jupyter notebook with active cells. Scroll down the page to see some nice examples.

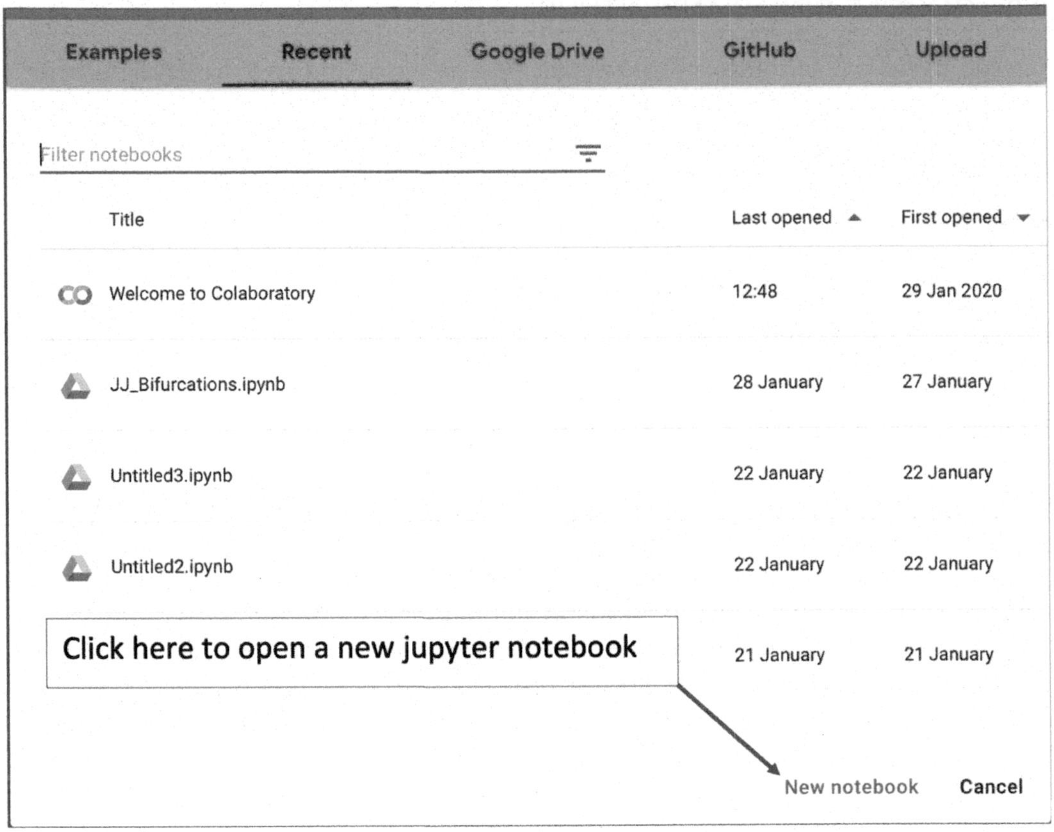

Figure 3.8 Login to Google Colab to get this web page. Open a previous notebook or click on **New notebook**.

In November 2021, GiHub reported to have over 73 million developers and more than 200 million repositories making it the largest source code host in the world. It is commonly used to host open-source projects. As far as the reader is concerned, it

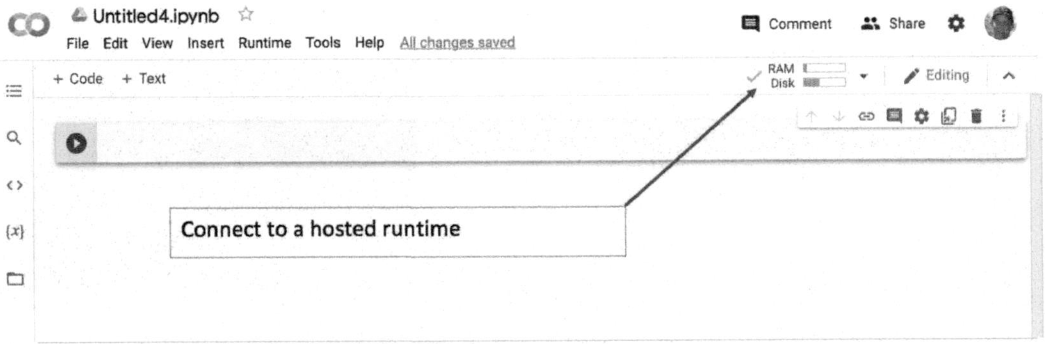

Figure 3.9 A jupyter notebook open through Google Colab. The user can now perform cloud computing by connecting to a hosted runtime.

allows you to work collaboratively with your partners on Python code. For a gentle introduction to GitHub, the reader is directed to [2].

EXERCISES

3.1 Use Python to solve the following differential equations:

(a) $\frac{dy}{dx} = -\frac{y}{x}$.

(b) $\frac{dy}{dx} = \frac{y}{x^2}$.

(c) $\frac{dx}{dt} + x^2 = 1$.

(d) $\frac{dx}{dt} + x = \sin(t)$.

(e) $\frac{d^2y}{dt^2} + 5\frac{dy}{dt} + 6y = 10\sin(t)$.

3.2 Use Python to animate the curve $y = x^3 + bx + 1$, for $-4 \le b \le 2$.

3.3 Animate the parametric curve defined by $x(t) = \cos(t)$, $y(t) = \sin(nt)$, for n from $n = 0$ to $n = 4$, when $0 \le t \le 2\pi$.

3.4 Use Python to generate an interactive plot of $y = Ae^{Bx} + C$, for $-2 \le A \le 2$, $-1 \le B \le 1$, and $-3 \le C \le 3$.

3.5 Use Python to generate an interactive parametric plot of $x(t) = A\sin(Bt)$, $y(t) = C\cos(Dt)$, for $-2 \le A, C \le 2$, $0 \le B, D \le 2\pi$, and $0 \le t \le 8\pi$.

Solutions to the Exercises may be found here:

https://drstephenlynch.github.io/webpages/Solutions_Section_1.html.

FURTHER READING

[1] Costa G.B. and Bronson R. (2014). *Differential Equations (SCHAUM), 4th Ed.* McGraw-Hill Education.

[2] Guthals, S. and Haack, P. (2019). *GitHub for Dummies.* For Dummies.

[3] Driscoll, M. (2018). *Jupyter Notebook 101.* Bowker.

## CHAPTER 4

Python for AS-Level (High School) Mathematics

High School students around the world study mathematics without much technology; however, more and more countries are now incorporating this into their syllabi. This is true for the new A-Level Mathematics syllabus introduced in the UK in 2017; unfortunately, this only extends to graphing calculators in this country. This chapter shows students and teachers of A-Level (High School) Mathematics how to use Python to gain a deeper understanding of the subject whilst at the same time learning to code. The author will refer to A-Level Maths only from now on, as the material in this chapter is taken from the UK syllabus. A-level Mathematics students and teachers should read Sections 2.2 and 2.3 before starting on this chapter. Section 3.2 shows how animations and interactive plots can be used to give a deeper understanding.

For the mathematical theory, see [1], and to access thousands of practice A-Level Mathematics questions, the reader is directed to [2]. For AS Level, see [3]. For High School Math, see [5] and [6]. Conrad Wolfram provides an education blueprint for the AI age, [7], proposing Mathematics as a computational subject.

Aims and Objectives:

- To use Python to aid in the study of AS-Level Mathematics.

At the end of the chapter, the reader should be able to

- solve problems from the AS-Level syllabus;

- use Python to check hand-written solutions;

- create new problems for added practice;

- gain a deeper understanding of mathematics.

DOI: 10.1201/9781003285816-4

The topics covered in this chapter include:

AS-Level Mathematics Topics

1. Indices and Surds
2. Quadratic Equations
3. Equations and Inequalities
4. Coordinate Geometry
5. Trigonometry
6. Polynomials
7. Graphs and Transformations
8. The Binomial Expansion
9. Differentiation
10. Integration
11. Vectors
12. Exponentials and Logs
13. Data Collection
14. Data Processing
15. Probability
16. The Binomial Distribution
17. Hypothesis Testing
18. Kinematics
19. Forces and Newton's Laws
20. Variable Acceleration
21. Proof

4.1 AS-LEVEL MATHEMATICS (PART 1)

Historically, in the first year of an A-Level Maths course, students cover AS-Level material, and in the second year, they move on to A-Level material. This first section covers material from the AS-Level syllabus by means of example.

1. Indices and Surds:
(a) Solve for x: $8^x = 0.8$.
(b) Simplify (i) $(2 + 2\sqrt{2})(-\sqrt{2} + 4)$; (ii) $\frac{6}{5\sqrt{2}}$; (iii) $(1 + \sqrt{3})^4$.

```
from sympy import *
x = symbols("x")
solve(8**x - 0.8 , x)
simplify((2 + 2 * sqrt(2))*(-sqrt(2) + 4))
simplify(6 / (5 * sqrt(2)))
expand((1 + sqrt(3))**4)
```

Solutions:
(a) $x = -0.107309364962454$; (b) (i) $4 + 6\sqrt{2}$; (ii) $\frac{3\sqrt{2}}{5}$; (iii) $16\sqrt{3} + 28$.

2. Quadratic Equations: Solve the quadratic equation: $3x^2 - 4x - 5 = 0$.

```
from sympy import *
x = symbols("x")
solve(3 * x**2 - 4 * x - 5 , x)
```

Solutions:
$\left[\frac{2}{3} + \frac{\sqrt{19}}{3}, \frac{2}{3} - \frac{\sqrt{19}}{3}\right].$

3. Simultaneous Equations and Inequalities:
(a) Solve the simultaneous equations: $2x + y = 4$ and $x^2 - y = 4$.
(b) Solve the inequality: $x^2 - 4x + 1 \geq x - 3$.

```
from sympy import *
x , y = symbols("x y")
solve((2 * x + y - 4 , x**2 - y - 4), x , y)
solve(x**2 - 4 * x + 1 >= x - 3 , x)
```

Solutions:
(a) $[(-4, 12), (2, 0)]$;   (b) $(4 \leq x \wedge x < \infty) \vee (x \leq 1 \wedge -\infty < x)$.

4. Coordinate Geometry:
Plot the circle $x^2 + y^2 - 2x - 2y - 23 = 0$ and the tangent line $4x - 3y - 26 = 0$ on one graph. See Figure 4.1.

```
import numpy as np
import matplotlib.pyplot as plt
# Set the dimensions of the figure.
fig = plt.figure(figsize = (6, 6))
x, y = np.mgrid[-6:8:100j,-6:8:100j]
z = x**2 + y**2 - 2 * x - 2 * y
# You can plot multiple contours if desired.
plt.contour(x, y, z, levels = [23])
x = np.linspace(2, 8, 100)
y = (4 * x - 26) / 3
plt.plot(x, y)
plt.xlabel("x")
plt.ylabel("y")
plt.show()
```

5. Trigonometry:
Plot the functions $y = 2\sin(3t)$ and $y = 3\cos(2t - \pi)$ on one graph. See Figure 4.2.

```
import numpy as np
import matplotlib.pyplot as plt
t = np.linspace(-2 * np.pi, 2 * np.pi, 100)
# LaTeX code in labels.
plt.plot(t, 2 * np.sin(3 * t), label = "$2\sin(3t)$")
plt.plot(t, 3 * np.cos(2 * t - np.pi), label ="$3\cos(2t-\pi)$")
legend = plt.legend()
plt.show()
```

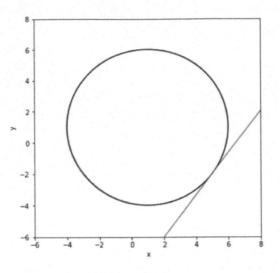

Figure 4.1 The circle $x^2 + y^2 - 2x - 2y - 23 = 0$, and the tangent line $4x - 3y - 26 = 0$.

6. Polynomials:
(a) Factorize: $2x^4 - x^3 - 4x^2 - 5x - 12$.
(b) Expand: $(x^2 - x + 1)(x^3 - x^2 + x)$.

```
from sympy import *
x = Symbol("x")
p1 = factor(2 * x**4 - x**3 - 4 * x**2 - 5 * x - 12)
p2 = expand((x**2 - x + 1) * (x**3 - x**2 + x))
print(p1)
print(p2)
```

Solutions:
$p1 = (2*x+3)(x**3-2*x**2+x-4)$; $p2 = x**5-2*x**4+3*x**3-2*x**2+x$.

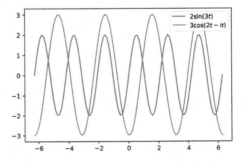

Figure 4.2 The functions $y = 2\sin(3t)$ and $y = 3\cos(2t - \pi)$.

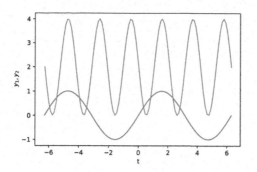

Figure 4.3 The functions $y = 2\sin(3t)$ and $y = 3\cos(2t - \pi)$.

7. Graphs and Transformations: Plot the functions $y_1 = 2sin(t)$ and $y_2 = 2 + 2\sin(3t + \pi)$ on one graph. See Figure 4.3.

```
import numpy as np
import matplotlib.pyplot as plt
t = np.linspace(-2 * np.pi, 2 * np.pi, 100)
plt.plot(t, np.sin(t), t, 2 + 2 * np.sin(3 * t + np.pi))
plt.xlabel("t")
plt.ylabel("$y_1, y_2$")
plt.show()
```

8. The Binomial Expansion:
(a) Expand $(1 - 2x)^6$.
(b) Plot the graph of $B(n) = \left(1 + \frac{1}{n}\right)^n$. See Figure 4.4.

```
from sympy import expand, Symbol
import numpy as np
import matplotlib.pyplot as plt
x = Symbol("x")
p1 = expand((1 - 2 * x)**6)
print("(1 - 2 * x)**6 =", p1)
n = np.linspace(1, 100, 100)
plt.plot((1 + 1/ n)**n)
plt.xlabel("n")
plt.ylabel("B(n)")
plt.show()
```

Solutions:
(a) $64*x**6 - 192*x**5 + 240*x**4 - 160*x**3 + 60*x**2 - 12*x + 1.$

9. Differentiation: Determine the first and second derivatives of $y = x^7 - 4x^5$.

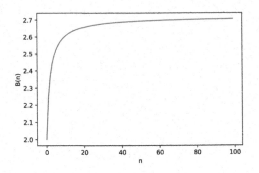

Figure 4.4 The function $B(n) = \left(1 + \frac{1}{n}\right)^n$.

```
from sympy import diff, Symbol
x = Symbol("x")
dy = diff(x**7 - 4 * x**5 , x)
print(dy)
d2y = diff(x**7 - 4 * x**5 , x , 2)
print(d2y)
```

Solutions:
$dy = 7*x**6 - 20*x**4; \quad d2y = 2*x**3*(21*x**2 - 40)$.

10. Integration:
(a) Integrate $I_1 = \int x^4 + 3x^3 - 8\,dx$.
(b) Integrate $I_2 = \int_{x=0}^{\pi} \sin(x)\cos(x)\,dx$.

```
from sympy import *
x = Symbol("x")
i1 = integrate(x**4 + 3 * x**3 - 8)
print("I_1 =", i1, "+ c")
i2 = integrate(sin(x) * cos(x), (x, 0, pi))
print("I_2 =",i2)
```

Solutions:
(a) $I_1 = x**5/5 + 3*x**4/4 - 8*x + c$; (b) $I_2 = 0$.

4.2 AS-LEVEL MATHEMATICS (PART 2)

This second section covers more material from the first year AS-Level syllabus by means of example.

11. Vectors: Given that $\underline{u} = (1, 3)$ and $\underline{v} = (3, -5)$, determine
(a) $\underline{w} = 2\underline{u} + 3\underline{v}$;
(b) the distance between $\underline{u}$ and $\underline{v}$.

```
from sympy import *
import numpy as np
u = np.array([1, 3])
v = np.array([3, -5])
w = 2 * u + 3 * v
print("w =", w)
d = np.linalg.norm(u - v)
print("norm(u-v) =", d)
```

Solutions:
(a) $w = [11, -9]$; (b) $\text{norm}(u - v) = 8.246211251235321$.

12. Exponentials and Logarithms:
(a) Determine e^{10} and $\log_{10}(22026.4658)$.
(b) Solve the equation $7^{x-3} = 4$.

```
import numpy as np
from sympy import *
exp10 = np.exp(10)
print("exp(10) =", exp10)
log10 = np.log(22026.4658)
print("log(22026.4658) =", log10)
x = Symbol("x")
sol = solve(7**(x-3) - 4)
print("The solution to", "7**{x-3}=4 is,", sol)
```

Solutions:
(a) $e^{10} = 22026.465794806718$, $\log(22026.4658) \approx 10.0$; (b) $x = \log(1372)/\log(7)$.

13. Data Collection:
(a) Select a list of 10 random numbers from a list of the first 25 natural numbers.
(b) Select a list of 10 random numbers from a list of the first 100 even natural numbers.

```
import math, random, statistics
a = list(range(1, 26))
data1 = random.sample(a, 10)
print("d1 =", data1)
b = list(range(2, 102, 2))
data2 = random.sample(b, 10)
print("d2 =", data2)
```

Solutions:
(a) $d1 = [22, 15, 21, 4, 5, 25, 9, 20, 19, 23]$; (b) $d2 = [88, 22, 10, 34, 100, 58, 56, 42, 48, 24]$.

14. Data Processing: Determine the mean, median, mode, variance and standard deviation of the data set, $[43, 39, 41, 48, 49, 56, 55, 44, 44, 32, 20, 44, 48, 2, 98]$.

```
import statistics as stats
data1 = [43, 39, 41, 48, 49, 56, 55, 44, 44, 32, 20, 44, 48, 2, 98]
mean_data1 = stats.mean(data1)
print("mean_data1 =", mean_data1)
median_data1 = stats.median(data1)
print("median_data1 =", median_data1)
mode_data1 = stats.mode(data1)
print("mode_data1 =", mode_data1)
var_data1 = stats.variance(data1)
print("var_data1 =", var_data1)
stdev_data1 = stats.stdev(data1)
print("stdev_data1 =", stdev_data1)
```

Solutions:
Mean = 44.2; median = 44; mode = 44; variance = 411.17; standard deviation = 20.28.

15. Probability: The following shows the results of a survey on types of exercises taken by a group of 100 pupils: 65 run, 48 swim, 60 cycle, 40 run and swim, 30 swim and cycle, 35 run and cycle and 25 do all 3. Plot a Venn diagram to represent these data. Let A denote the set of pupils who run, B represent the set of pupils who swim and C contain the set of pupils who cycle. See Figure 4.5.

```
# Venn Diagram.
# Install the matplotlib-venn library on to your computer.
# In the Terminal type: conda install -c conda-forge matplotlib-venn
# Alternatively, run the code in Google Colab.
import matplotlib.pyplot as plt
import numpy as np
from matplotlib_venn import venn3
fig, ax = plt.subplots(facecolor="lightgray")
ax.axis([0, 1, 0, 1])
v = venn3(subsets = (15, 3, 15, 20, 10, 5, 25))
ax.text(1, 1, "7")
plt.show()
```

16. The Binomial Distribution: The binomial distribution is an example of a discrete probability distribution function.

Given $X \sim B(n, p)$, then $P(X = r) = \binom{n}{r} p^r (1 - p)^{n-r}$, where X is a binomial random variable, and p is the probability of r success from n independent trials. Note

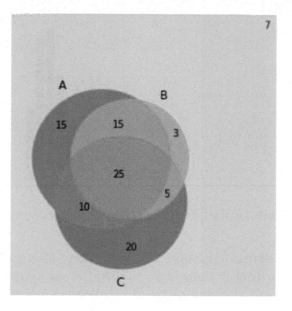

Figure 4.5 Venn diagram.

that (i) the expectation, $E(X) = np$; (ii) the variance, $\text{var}(X) = np(1 - p)$ and (iii) the standard deviation, $\text{std} = \sqrt{np(1 - p)}$.

Manufacturers of a bag of sweets claim that there is a 90% chance that the bag contains some toffees. If 20 bags are chosen, what is the probability that:
(a) All bags contain toffees?
(b) More than 18 bags contain toffees?

```
from scipy.stats import binom
import matplotlib.pyplot as plt
n , p = 20 , 0.9
r_values = list(range(n+1))
dist = [binom.pmf(r , n , p) for r in r_values]
plt.bar(r_values , dist)
plt.show()
print("Mean = ", binom.mean(n = 20 , p = 0.9))
print("Variance = ", binom.var(n = 20 , p = 0.9))
print("Standard Deviation = ", binom.std(n = 20 , p = 0.9))

PXis20 = binom.pmf(20, n, p)        # Probability mass function.
print("P(X=20)=" , PXis20)
PXGT18 = 1 - binom.cdf(18, n, p)  # Cumulative distribution function.
print("P(X>18)=" , PXGT18 )         # P(X>18)=1-P(X<=18).
```

Solutions:
Mean $= 18$; Variance $= 11.8$; Standard Deviation $= 11.3416$. (a) $P(X = 20) = 0.1216$;
(b) $P(X > 18) = 0.3917$. Note the use of the cumulative distribution function (CDF).

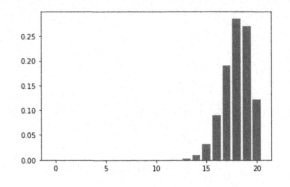

Figure 4.6 Bar graph of a binomial distribution.

17. Hypothesis Testing: Example: The probability of a patient recovering from illness after taking a drug is thought to be 0.85. From a trial of 30 patients with the illness, 29 recovered after taking the drug. Test, at the 10% level, whether the probability is different to 0.85.

So the null-hypothesis is, $H_0 = 0.85$,

$H_1 \neq 0.85$. (Two-tail test)

The program below gives $P(X \geq 29) = 0.0480 < 0.05$. Therefore, the result is significant and so we reject H_0. There is evidence to suggest that the probability of recovery from the illness after taking the drug is different to 0.85. See Figure 4.7.

```
from scipy.stats import binom
import matplotlib.pyplot as plt
n , p = 30 , 0.85
r_values = list(range(n+1))
dist = [binom.pmf(r , n , p) for r in r_values]
plt.bar(r_values , dist)
plt.show()
print("Mean = ", binom.mean(n = 30 , p = 0.85))
print("Variance = ", binom.var(n = 30 , p = 0.85))
print("Standard Deviation = ", binom.std(n = 30 , p = 0.85))
PXGTE29 = 1 - binom.cdf(28, n, p)  # Cumulative distribution function.
print("P(X>=29)=" , PXGTE29 )      # P(X>=29)=1-P(X<=28).
```

Solution:
Mean $= 25.5$; Variance $= 3.825$; Standard Deviation $= 1.9558$; $P(X \geq 29) = 0.04803$.

18. Kinematics: Plot a velocity-time graph. See Figure 4.8.

```
import matplotlib.pyplot as plt
xs = [0, 3, 9, 15]
```

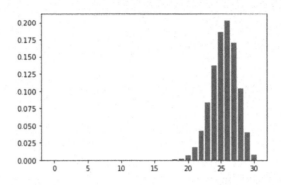

Figure 4.7 A bar chart of the data relating to patients recovering from an illness after taking a drug.

```
ys = [14, 8, 8, -4]
plt.axis([0, 15, -4, 14])
plt.plot(xs, ys)
plt.xlabel("time(s)")
plt.ylabel("velocity (m/s)")
plt.show()
```

19. Forces and Newton's Laws: Plot a forces diagram. See Figure 4.9.

```
import matplotlib.pyplot as plt
import matplotlib.pyplot as plt
ax = plt.axes()
plt.axis([-2, 2, -2, 2])
ax.arrow(0,0,0,1.5,head_width=0.05,head_length=0.1,fc="k",ec="k")
ax.arrow(0,0,1.5,0,head_width=0.05,head_length=0.1,fc="r",ec="r")
ax.arrow(0,0,-1.5,-1.5,head_width=0.05,head_length=0.1,fc="b",ec="b")
plt.text(0.1,1.5,"$F_1$",fontsize=15)
plt.text(1.5,0.1,"$F_2$",{"color":"red","fontsize":15})
```

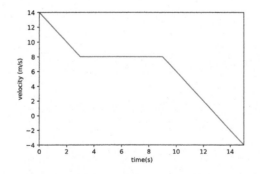

Figure 4.8 Plotting data points for a velocity-time graph.

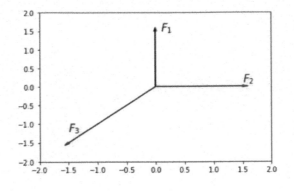

Figure 4.9 A forces diagram.

```
plt.text(-1.5,-1.2,"$F_3$",{"color":"blue","fontsize":15})
plt.show()
```

20. Variable Acceleration: Plot a velocity-time graph using an array. See Figure 4.10.

```
import numpy as np
import matplotlib.pyplot as plt
pts = np.array([0,1,2,3,3,3,2,1,3,6,10,15])
fig = plt.figure()
plt.plot(pts)
plt.xlabel("time(s)")
plt.ylabel("velocity (m/s)")
plt.show()
```

21. Proof:
(a) For the natural numbers up to 100, investigate when $2n^2$, is close to a perfect square. This will give a good approximation to $\sqrt{2}$. Can you explain why?

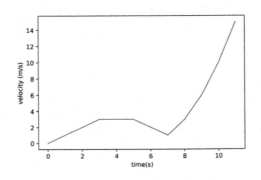

Figure 4.10 Using an array to plot a velocity-time graph.

(b) Prove by elimination that there is only one solution to the equation: $m^2 - 17n^2 = 1$, where m and n are natural numbers and, $1 \le m \le 500, 1 \le n \le 500$.

```
# Part (a).
import numpy as np
for n in range(1, 101):
    num1 = np.sqrt(2 * n**2)
    num2 = int(np.round(np.sqrt(2 * n**2)))
    if abs(num1 - num2) < 0.01:
        print(num2, "divided by", n, "is approximately sqrt(2)")

# Part (b).
# An example of a program with for loops and if statements.
# The final value, 501, is not included in range in Python.
for m in range(1, 501):
    for n in range(1, 501):
        if m**2 - 17 * n**2 == 1:          # Note the Boolean ==.
            print("The unique solution is, m =",m, "and n =",n)
```

Solutions:
(a) Approximations to $\sqrt{2}$ are $\frac{99}{7}$ and $\frac{140}{99}$; (b) The unique solution is, $m = 33, n = 8$.

At this point, it is worth pointing out that there are mathematical research groups around the world who are attempting mathematical proof using AI. See the recent Nature article entitled "Advancing mathematics by guiding human intuition with AI" [4], for example. AI is discussed in some detail in the third section of this book.

EXERCISES

4.1 Use Python to:

(a) Simplify $(2 - \sqrt{5})^4$.

(b) Solve $2x^2 - 3x - 2 = 0$.

(c) Solve the simultaneous equations, $y - x = 1, x^2 + y^2 = 2$.

(d) Determine the equation of the line tangent to the circle $x^2 + y^2 - 10x - 10y + 45 = 0$, at the point $P(3, 4)$. Plot the curves on one graph.

(e) Plot the curve $y = 2\sin(2\pi t)$.

4.2 Use Python to:

(a) Factorize $x^4 - 2x^3 - 6x^2 + 6x + 9$.

(b) Plot $y = \cos(x)$ and $y = 1 - 2\cos(x - \pi)$ on the same graph.

(c) Expand $(1 - 2x)^5$.

(d) Determine the first and second derivatives of $y = 4x^8 - 3x^2$.

(e) Find $I = \int_{x=0}^{3} 3x^{\frac{1}{2}} - x^{\frac{3}{2}} \, dx$.

4.3 Using Python:

(a) Given that $\underline{u} = (2, -3)$, $\underline{v} = (1, 4)$, determine $\underline{w} = 5\underline{u} - 6\underline{v}$.

(b) Solve the equation $5^{2x+3} = 10$.

(c) Print a list of 20 random numbers from a list of the first 500 natural numbers.

(d) Determine the mean, median, mode, variance and standard deviation from the list: $[2, 3, 5, 67, 46, 34, 34, 18, 4, 54, 24, 34, 56, 35, 66]$.

(e) Plot a Venn diagram representing the following data. Four people cycle to work, 5 both cycle and take the train, 3 travel by train, 7 walk and 6 don't travel by cycle, train or walking."

4.4 Use Python to answer these questions.

(a) Plot a probability mass function for a binomial distribution when $n = 25$ and $p = 0.3$. Compute the mean, variance and standard deviation.

(b) When Thalia buys lunch at school, there is a probability of 0.6 that the canteen has her favorite sandwich in stock. After only one sandwich being in stock in the last five days, she is certain the probability of sandwiches being in stock has decreased. Test, at the 5% significance level, if Thalia is correct.

(c) Plot a velocity-time graph with data points: $xs = [0, 4, 6, 10, 16]$ and $ys = [2, 6, 2, 14, 18]$.

(d) The forces $\mathbf{F_1} = 2\mathbf{i} + 3\mathbf{j}$, $\mathbf{F_2} = 4\mathbf{i} - \mathbf{j}$, $\mathbf{F_3} = -3\mathbf{i} + 2\mathbf{j}$ and $\mathbf{F_4} = a\mathbf{i} + b\mathbf{j}$, act on a particle which is in equilibrium. Determine the values of a and b and plot a diagram illustrating the forces acting on the particle.

(e) Plot a velocity-time graph using the array:
`[0 2 3 4 4 4 6 1 3 5 7 7 1 0]`.

(f) Prove that $(2n + 3)^2 - (2n - 3)^2$ is a multiple of 6 for $0 \leq n \leq 100$. Can you prove it is true for all n?

Solutions to the Exercises may be found here:

`https://drstephenlynch.github.io/webpages/Solutions_Section_1.html`.

FURTHER READING

[1] Attwood, G. (2017). *Edexcel AS and A Level Mathematics, Statistics and Mechanics.* Pearson Education.

[2] CGP Books. (2021). *A-Level Maths Textbook: Year 1 and 2: thousands of practice questions for the full course.* Coordination Group Publications Ltd.

[3] CGP Books. (2017). *A-Level Maths for AQA: Year 1 and AS Complete Revision and Practice with Online Edition.* Coordination Group Publications Ltd.

[4] Davies, A., Velickovic, P., Buesing, L. et al. (2021). Advancing mathematics by buiding human intuition with AI. *Nature* **600**, 70–74.

[5] Dillon, F.L. andf Martin W.G. (2017). *The Common Core Mathematics Companion.* Corwin.

[6] Martin, J. (2020). *High School Math Made Understandable Book 3: Math 9, 10, 11 and 12.* Independently Published.

[7] Wolfram, C. (2020). *The Math(s) Fix: An Education Blueprint for the AI Age.* Wolfram Media.

## CHAPTER 5

Python for A-Level (High School) Mathematics

A-Level Mathematics is currently the most popular High School subject in the UK. For theory and examples, please see [1] and [2].

Aims and Objectives:

- To use Python to aid in the study of A-Level Mathematics.

At the end of the chapter, the reader should be able to

- solve problems from the A-Level syllabus;

- use Python to check hand-written solutions;

- create new problems for added practice;

- gain a deeper understanding of mathematics.

The topics covered in this chapter include:

A-Level Mathematics Topics

22. Trigonometry and Series

23. Sequences and Series

24. Functions

25. Differentiation

26. Trigonometric Functions

27. Algebra

28. Trigonometric Identities

29. Further Differentiation

30. Further Integration

31. Parametric Equations

32. Vectors in 3D

33. Differential Equations

34. Numerical Methods

35. Probability

36. Probability Distributions

37. Hypothesis Testing

DOI: 10.1201/9781003285816-5

38. Kinematics (SUVAT) 39. Forces and Motion

40. Moments 41. Projectiles

42. Friction

5.1 A-LEVEL MATHEMATICS (PART 1)

This first section covers material from the second year A-Level syllabus by means of example.

22. Trigonometry and Series: See Figure 5.1. By plotting curves, show that:
(a) $\sin(t) \approx t$, near $t = 0$;
(b) $\cos(t) \approx 1 - \frac{t^2}{2}$, near $t = 0$.

```python
import numpy as np
import matplotlib.pyplot as plt
plt.figure(1)
plt.subplot(121)
t = np.linspace(- 2, 2, 100)
plt.plot(t, np.sin(t), t, t)
plt.xlabel("t")
plt.ylabel("f(t)")
plt.subplot(122)
t = np.linspace(- 2, 2, 100)
plt.plot(t, np.cos(t), t, 1 - t**2 / 2)
plt.xlabel("t")
plt.ylabel("f(t)")
plt.subplots_adjust(wspace=1)
plt.show()
```

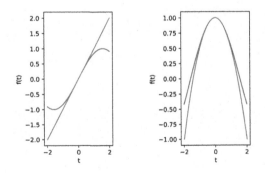

Figure 5.1 Approximate polynomials of $\sin(t)$ and $\cos(t)$ near $t = 0$.

23. Sequences and Series: An arithmetic sequence, $a_n = a + (n-1)d$, has sum

$$S_n = \sum_{i=1}^{n} a_n = \frac{n}{2} \left(2a + (n-1)d \right),$$

and a geometric sequence $g_n = ar^{n-1}$, has sum

$$S_n = \sum_{i=1}^{n} g_n = \frac{a \left(r^n - 1 \right)}{r - 1}, r \neq 1.$$

Find:
(a) a_{20} and S_{20}, when $a = 3$ and $d = 6$;
(b) g_{30} and S_{30}, when $a = 4$ and $r = 2$.

```python
# Arithmetic Series
def nthterm(a, d, n):
    term = a + (n - 1) * d
    return(term)
def sumofAP(a, d, n):
    sum, i = 0, 0
    while i < n:
        sum += a
        a = a + d
        i += 1                  # i = i + 1.
    return sum
a, d, n = 3, 6, 20
print("The", n, "th", "term is", nthterm(a, d, n))
print("The sum of the", n, "terms is", sumofAP(a, d, n))

# Geometric Series, given a=4 and r=2.
print("g(30)= ", 4 * 2**29)
print("S(30)= ", 4 * (2**30 -  1) / (2 - 1))
```

Solutions:
(a) $a_{20} = 117$, $S_{20} = 1200$; (b) $g_{30} = 2147483648$, $S_{30} = 4294967292$.

24. Functions: Given that $f(x) = 1 - x^2$, determine:
(a) $f(2)$;
(b) $f(f(f(x)))$.

```python
from sympy import expand, Symbol
x = Symbol("x")
def f(x):
    return(1 - x**2)
```

```
print("f(x) = ", f(x))
print("f(2) = ", f(2))
print("f(f(f(x))) = ", expand(f(f(f(x)))))
```

Solutions:
(a) $f(2) = -3$; (b) $f(f(f(x))) = -x**8 + 4*x**6 - 4*x**4 + 1$.

25. Differentiation: Given that $f(x) = \frac{x}{\sqrt{2x-1}}$, determine $\frac{df}{dx}\big|_{x=5}$.

```
from sympy import diff, sqrt, Symbol
x = Symbol("x")
df = diff(x / sqrt(2 * x - 1), x)
print("df/dx = ", df)
dfx = df.subs(x, 5)
print("Gradient at x = 5 is", dfx)
```

Solutions:
(a) $\frac{df}{dx} = -x/(2*x - 1)**(3/2) + 1/sqrt(2*x - 1)$; (b) $\frac{df}{dx}\big|_{x=5} = \frac{4}{27}$.

26. Trigonometric Functions:
(a) Plot the curve $\sec(t) = \frac{1}{\cos(t)}$. See Figure 5.2.
(b) Solve the trigonometric equation, $2\cos(x) + 5\sin(x) = 3$.

```
# Part (a).
import numpy as np
import matplotlib.pyplot as plt
t = np.linspace(- 2 * np.pi, 2 * np.pi, 1000)
plt.axis([-2 * np.pi, 2 * np.pi, -10, 10])
plt.plot(t, 1 / np.cos(t))
plt.xlabel("t")
plt.ylabel("sec(t)")
plt.show()

# Part (b).
from sympy import *
x = Symbol("x")
solve(2 * cos(x) + 5 * sin(x) - 3)
```

Solutions:
(b) $[2*atan(1 - 2*sqrt(5)/5), 2*atan(2*sqrt(5)/5 + 1)]$.

27. Algebra:.
(a) Split $\frac{(2x+3)}{(x-2)(x-3)(x-4)}$ into partial fractions;
(b) Determine the $\mathcal{O}(x^{10})$ Taylor series expansion of $\frac{1}{2-x}$, at $x = 0$;

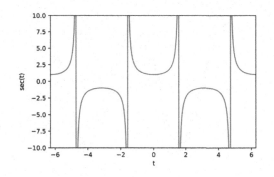

Figure 5.2   A graph of the secant curve.

```
from sympy import *
x = Symbol("x")
apart((2 * x + 3)/((x - 2) * (x - 3) * (x - 4)))

x = Symbol("x")
Taylor_series = series(1 / (2 - x) , x , 0 , 10)
print("The Taylor series of 1/(2-x) is ", Taylor_series)
```

Solutions:
(a) $\frac{7}{2(x-2)} - \frac{9}{x-3} + \frac{11}{2(x-4)}$; (b) $1/2 + x/4 + x**2/8 + x**3/16 + x**4/32 + x**5/64 + x**6/128 + x**7/256 + x**8/512 + x**9/1024 + O(x**10)$.

28. Trigonometric Identities: See Figure 5.3. Show that the curves $f(t) = \sqrt{2}\cos(t) + \sin(t)$ and $g(t) = \sqrt{3}\cos(t - \alpha)$, are equivalent, where $\alpha = \tan^{-1}\left(\frac{1}{\sqrt{2}}\right)$.

```
import numpy as np
from sympy import sqrt
from math import atan
import matplotlib.pyplot as plt
alpha = atan(1/sqrt(2))
t = np.linspace(- 2 * np.pi, 2 * np.pi, 100)
plt.plot(t, sqrt(2) * np.cos(t) + np.sin(t), "r-",linewidth = 10)
plt.plot(t, sqrt(3) * np.cos(t - alpha), "b-", linewidth = 4)
plt.xlabel("t")
plt.ylabel("f(t)")
plt.show()
```

29. Further Differentiation: The rules of differentiation at A-Level are:
Product rule:

$$\frac{d}{dx}(uv) = u\frac{dv}{dx} + v\frac{du}{dx}.$$

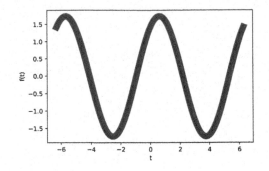

Figure 5.3 A graph of $f(t)$, red curve, and $g(t)$, blue curve.

Chain rule:

$$\frac{d}{dx}(g(u(x))) = \frac{dg}{du}\frac{du}{dx}.$$

Quotient rule:

$$\frac{d}{dx}\left(\frac{u}{v}\right) = \frac{vdu - udv}{v^2}.$$

Differentiate:
(a) $f(x) = 3x^2 e^{-3x}$;
(b) $g(x) = \cos(3x^3 - x)$;
(c) $h(x) = \frac{3\sin(x)}{\ln(7x)}$.

```
from sympy import diff, exp, log, sin, cos, Symbol
x = Symbol("x")
print("df = ", diff(3 * x**2 * exp(- 3 * x) , x))
print("dg = ", diff(cos(3 * x**3 - x) , x))
print("dh = ", diff(3 * sin(x) / log(7 * x) , x))
```

Solutions:
(a) $df = -9 * x * *2 * \exp(-3 * x) + 6 * x * \exp(-3 * x)$; (b) $dg = -(9 * x * *2 - 1) * \sin(3 * x * *3 - x)$; (c) $dh = 3 * \cos(x)/\log(7 * x) - 3 * \sin(x)/(x * \log(7 * x) * *2)$.

30. Further Integration: The rules of integration at A-Level are:
Substitution:

$$\int f(u(x))u'(x)dx = \int f(u)du.$$

By parts:

$$\int udv = uv - \int vdu.$$

Integrate:
(a) $I_1 = \int (2x^3 + 1)^7 x^2 \, dx$;
(b) $I_2 = \int x \cos(3x) \, dx$.

```
from sympy import *
x = Symbol("x")
I_1 = integrate((2 * x**3 + 1)**7 * x**2, x)

print("I_1=",I_1, "+ c")
I_2 = integrate(x * cos(3 * x), x)
print("I_2=",I_2, "+ k")
```

Solutions:
(a) $I_1 = 16*x**24/3 + 64*x**21/3 + 112*x**18/3 + 112*x**15/3 + 70*x**12/3 + 28*x**9/3 + 7*x**6/3 + x**3/3 + c$; (b) $I_2 = x*sin(3*x)/3 + cos(3*x)/9 + k$.

31. Parametric Equations: See Figure 5.4. Plot the parametric curve defined by the equations, $x(t) = 1 - \cos(t)$, $y(t) = \sin(2t)$, for $0 \le t \le 2\pi$.

```
import numpy as np
import matplotlib.pyplot as plt
t = np.linspace(0, 2 * np.pi, 100)
x = 1 - np.cos(t)
y = np.sin(2 * t)
plt.plot(x, y)
plt.xlabel("x")
plt.ylabel("y")
plt.show()
```

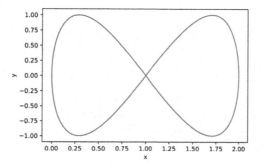

Figure 5.4   A parametric curve.

---

32. Vectors in 3D: Given $\underline{u} = (1, -1, 4)$ and $\underline{v} = (2, -1, 9)$, determine:
(a) $5\underline{u} - 6\underline{v}$;
(b) $\underline{u}.\underline{v}$;
(c) $\underline{u} \times \underline{v}$;
(d) the angle between $\underline{u}$ and $\underline{v}$.

```python
import numpy as np
from math import acos
u = np.array([1, -1, 4])
v = np.array([2, -1, 9])
w = 5 * u - 6 * v
print("w =", w)
print("dot(u,v)=", np.dot(u , v))
print("cross(u,v)=", np.cross(u , v))
theta = acos(np.dot(u,v)/(np.linalg.norm(u) * np.linalg.norm(v)))
print("theta =", theta, "radians")
```

Solutions:
(a) $[-7 \ 1 \ -34]$; (b) $\mathrm{dot}(u, v) = 39$; (c) $\mathrm{cross}(u, v) = [-5 \ -1 \ 1]$; (d) $\theta = 0.1325$ radians.

5.2   A-LEVEL MATHEMATICS (PART 2)

This second section covers more material from the second year A-Level syllabus by means of example.

33. Differential Equations: See Figure 5.5. Simple differential equations (ODEs) were introduced in Section 3.1. Here we show how to plot a solution curve to an initial value problem (IVP). Other ODEs will be solved later in the book.

The population of fish, $P(t)$ say, in a large lake is modeled using the ODE:

$$\frac{dP}{dt} = P(1 - 0.001P),$$

where $P(0) = 80 \ (\times 10^6)$. Plot a solution curve for $0 \le t \le 10$ (years).

```python
import numpy as np
from sympy import dsolve, Function, symbols
import matplotlib.pyplot as plt
P = Function("P")
t = symbols("t")
ODE = P(t).diff(t) - P(t) * (1 - 0.01 * P(t))
print("P(t) =", dsolve(ODE, ics = {P(0):80}))
```

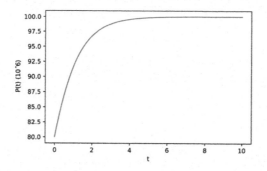

Figure 5.5 Solution curve showing the population of fish in a large lake. A physical interpretation is that initially there are 80×10^6 fish in the lake, and this population slowly increases to 100×10^6 (the carrying capacity of the lake) over a 10 year time interval.

```
t = np.linspace(0, 10, 100)
# Plot the solution to the ODE, P(t).
plt.plot(t, 100.0/(1.0  + 0.25*np.exp(-t)))
plt.xlabel("t")
plt.ylabel("P(t) $(10^6)$")
plt.show()
```

34. Numerical Methods: Use the Newton-Raphson method to find the root of $x^3 - 0.9x^2 + 2 = 0$, starting with the point $x_0 = 2$. Work to 8 decimal places throughout. Recall that:

$$x_{n+1} = x_n - \frac{f(x_n)}{f'(x_n)},$$

where $f'(x)$ is the derivative of $f(x)$.

```
def fn(x):
  return x**3 - 0.9 * x**2 + 2
def dfn(x):
  return 3 * x**2 - 1.8 * x
def NewtonRaphson(x):
  i=0
  h = round(fn(x) / dfn(x) , 8)
  while abs(h) >= 0.0001:
    h = round(fn(x) / dfn(x) , 8)
    x = x - h
    i += 1
    print("x(", i ,")=", round(x , 8))
# Start at x( 0 ) = 2.
NewtonRaphson(2)
```

Solutions:
x(1) is 1.23809524
x(2) is 0.17556905
x(3) is 9.02219202
x(4) is 6.11314524

⋮

x(19) is -1.07224453
x(20) is -1.02251451
x(21) is -1.02049367
x(22) is -1.02049041

35. Probability: See Figure 5.6. Generate some random data in Python and compare the histogram to a normal probability distribution function. Determine the mean and standard deviation of the data.

```
from scipy import stats
import numpy as np
import matplotlib.pyplot as plt
data = 50 * np.random.rand() * np.random.normal(10, 10, 100) + 20
plt.hist(data, density = True)
xt = plt.xticks()[0]
xmin, xmax = min(xt), max(xt)
lnspc = np.linspace(xmin, xmax, len(data))
m, s = stats.norm.fit(data)
pdf_g = stats.norm.pdf(lnspc, m, s)
plt.plot(lnspc, pdf_g)
print("Mean =", m)
print("Standard Deviation =", s)
plt.show()
```

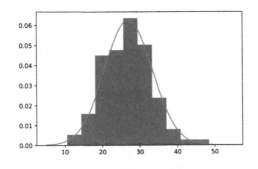

Figure 5.6 A histogram of random data and a probability density function.

36. Probability Distributions: The normal probability density function (PDF) is an example of a continuous PDF. The general form of a normal distribution is:

$$f(x) = \frac{1}{\sigma\sqrt{2\pi}} e^{-\frac{1}{2}\left(\frac{x-\mu}{\sigma}\right)^2},$$

where μ is the mean or expectation of the distribution, σ^2, is the variance, and σ is the standard deviation. The standard normal distribution, also known as the z-distribution, is a special normal distribution having a mean of zero and a standard deviation of one. It allows you to compare scores on different distributions, normalize scores for decision making, and find if the probability of a sample mean significantly differs from the population mean, for example.

Suppose that X is normally distributed with a mean of 80 and a standard deviation of 4. So, $X \sim N(80, 16)$. Plot the bell curve and determine:
(a) $P(X < 75)$;
(b) $P(X > 83)$;
(c) $P(72 < X < 76)$.

```python
import matplotlib.pyplot as plt
import numpy as np
import scipy.stats as stats
import math


mu , variance = 80 , 4**2
sigma = math.sqrt(variance)
x = np.linspace(mu - 3*sigma, mu + 3*sigma, 100)
plt.plot(x, stats.norm.pdf(x, mu, sigma))
plt.show()


sola = stats.norm.cdf(75, loc=mu, scale=sigma)
print("P(X<75)=", sola)
solb = 1 - stats.norm.cdf(83, loc=mu, scale=sigma)
print("P(X>83)=", solb)
solc = stats.norm.cdf(76, loc=mu, scale=sigma) - \
       stats.norm.cdf(72, loc=mu, scale=sigma)
print("P(72<X<76)=", solc)
```

Solutions:
(a) $P(X < 75) = 0.1056$; $P(X > 83) = 0.2266$; (c) $P(72 < X < 76) = 0.1359$.

37. Hypothesis Testing: The diameters of circular cardboard drinks packs produced by a certain machine are normally distributed with a mean of 9cm and standard deviation of 0.15cm. After the machine is serviced a random sample of 30 mats is selected and their diameters are measured to see if the mean diameter has altered. The mean of the sample was 8.95cm. Test at the 5% level, whether there is significant

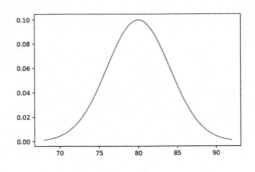

Figure 5.7 The normal probability density function.

evidence of a change in the mean diameter of mats produced by the machine.

Now, the null-hypothesis is, $H_0 : \mu = 9$,

The alternative hypothesis is, $H_1 : \mu \neq 9$. (Two-tailed test)

See below. Since $P(X <= 8.95) = 0.0339 > 0.025$, accept H_0.

There is not enough evidence to suggest there has been a change in the mean diameter of the mats.

```
import matplotlib.pyplot as plt
import numpy as np
import scipy.stats as stats
import math

mu , variance = 9 , 0.15**2 / 30
sigma = math.sqrt(variance)
x = np.linspace(mu - 3*sigma, mu + 3*sigma, 100)
plt.plot(x, stats.norm.pdf(x, mu, sigma))
plt.show()
sol = stats.norm.cdf(8.95, loc=mu, scale=sigma)
print("P(X<=8.95)=", sol)
```

Solution:
$P(X \leq 8.95) = 0.0339$.

38. Kinematics (SUVAT): Recall the SUVAT equations:

$$\underline{v} = \underline{u} + \underline{a}t,$$

$$\underline{v}^2 = \underline{u}^2 + 2\underline{a}\,\underline{s},$$

$$s = \underline{u}t + \frac{1}{2}\underline{a}t^2,$$

$$s = (\underline{u} + \underline{v})t/2,$$

where $\underline{s}$ is distance travelled, $\underline{u}$ is initial velocity, $\underline{v}$ is velocity, $\underline{a}$ is acceleration and t is time.

Given that, $\underline{u} = 2\underline{i} + 9\underline{j}$ ms^{-1}, $\underline{a} = -9.8\underline{j}$ ms^{-2} and $t = 0.2$ s, find the distance travelled.

```python
import numpy as np
u = np.array([2, 9])
a = np.array([0, -9.8])
t = 0.2
s = u * t + a * t**2 / 2
print("Distance travelled =", s, "m")
```

Solution:
(a) Distance travelled is [0.4 1.604] m.

39. Forces and Motion: Determine the resultant force given, $F_A = -20\underline{j}$ N and $F_B = -16\underline{i} - 16\underline{j}$ N.

```python
import numpy as np
FAx , FAy = 0 , -20
FBx = -16 * np.sin(np.radians(40.8))
FBy = -16 * np.cos(np.radians(40.8))
FR = np.array([FAx + FBx, FAy + FBy])
FR_norm = np.linalg.norm(FR)
FR_angle = np.degrees(np.arctan(FR[1] / FR[0]))
print("FR =", FR_norm, "N")
print("FR Angle =", FR_angle, "degrees.")
```

Solution:
$F_R = 33.77095$ N; F_R angle $= 71.96622$ degrees.

40. Moments: See Figure 5.7. Given that $F_1 = 5$ N, $F_2 = 2$ N, where would a force of $F_3 = 3$ N, need to be placed to balance the beam?

```python
import matplotlib.pyplot as plt
ax = plt.axes()
plt.plot([-10, 10], [0, 0])
plt.axis([-11, 11, -10, 10])
plt.plot([0], marker = "d")
ax.arrow(-5, 0, 0, -4, head_width = 1, head_length = 1, fc = "k", ec = "k")
ax.arrow(2.5, 0, 0, -1, head_width = 1, head_length = 1, fc = "r", ec = "r")
```

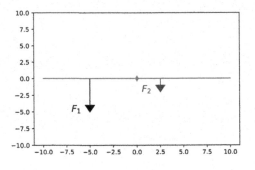

Figure 5.8 Forces on a beam.

```
plt.text(-7, -5, "$F_1$", fontsize = 15)
plt.text(0.5, -2, "$F_2$", {"color": "red", "fontsize":15})
print("x=", (5 * 5 - 2 * 2.5) / 3, "m")
plt.show()
```

41. Projectiles: See Figure 5.9. Plot the trajectory of a ball thrown from the top of a cliff. The ball is thrown from the top of a cliff of height 10 metres with initial velocity $\underline{u} = 20\sqrt{3}\underline{i} + 20\underline{j}\,\text{ms}^{-1}$.

```
import numpy as np
import matplotlib.pyplot as plt
x0, y0, sy, g, dt, t = 0, 10, 0, 9.8, 0.01, 0
vx0 = 20 * np.sqrt(3)
vy0 = 20
while sy >= 0:
    sx = vx0 * t
    sy = vy0 * t - g * t**2 / 2 + y0
    t = t + dt
    plt.plot(sx, sy, "r.")
```

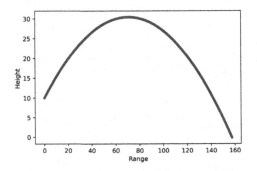

Figure 5.9 The trajectory of a ball thrown from the top of a cliff.

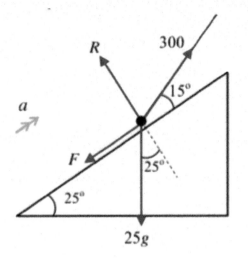

Figure 5.10   Forces diagram on a slope.

```
plt.xlabel("Range")
plt.ylabel("Height")
plt.show()
```

42. Friction: See Figure 5.10. Show that the acceleration of the particle up the slope is $6.005454\,\text{ms}^{-2}$, given that the coefficient of friction, μ say, is $\mu = \frac{1}{4}$.

```
import numpy as np
# Resolve forces normal to slope.
R = 25 * 9.8 * np.cos(np.radians(25)) - 300 * np.sin(np.radians(15))
print("R =", R, "N")
# Since particle is moving.
F = (1 / 4) * R
print("F =", F, "N")
# Resolve forces parallel to slope.
a_num = 300 * np.cos(np.radians(15)) - F - 25 * 9.8 * \
np.sin(np.radians(25))
a = a_num / 25
print("a =", a, "m/s/s.")
```

EXERCISES

5.1 Using Python:

(a) Show that $\sec(x) \approx 1 + \frac{x^2}{2} + \frac{5x^4}{24}$.

(b) Given $a = 2$ and $r = 3$, determine the 50th term of the geometric sequence and the sum of the first 50 terms.

(c) Given that $f(x) = 2x^2 - 1$ and $g(x) = x^3 + 1$, determine $g(f(-1))$.

(d) Given that $f(x) = \frac{2x}{x^2+3}$, determine $\frac{df}{dx}\big|_{x=2}$.

(e) Solve the trigonometric equation: $3\sin(x) + 4\cos(x) = 5$, for $0 \le x \le 2\pi$.

5.2 Use Python to:

(a) Split $\frac{3x}{(x+1)(x-1)(2x-5)}$ into partial fractions.

(b) Show that the curves $f(t) = \tan(2t)$ and $g(t) = \frac{2\tan(t)}{1-\tan^2(t)}$, are equivalent.

(c) Differentiate $y = \frac{xe^{\sin(x)}}{\cos(x)}$.

(d) Integrate $I = \int x^3 \sin(3x)\,dx$.

(e) Plot the parametric curve C with parametric equations, $x(t) = 4\cos\left(t + \frac{\pi}{6}\right), y(t) = 2\sin(t), 0 \le t < 2\pi$.

5.3 Using Python:

(a) Given that $\underline{u} = (-1, 2, 3)$ and $\underline{v} = (9, 0, -2)$, find the magnitude of the vector $2\underline{u} - 3\underline{v}$.

(b) The amount of drug in a patients body is N mg after t hours. An ODE to model this system is given by:

$$\frac{dN}{dt} = -\frac{1}{8}N(t+1).$$

Initially, there are 30 mg in a patient's body. Find the amount of drug in the body after 8 hours and plot a solution curve for this problem.

(c) Use the Newton–Raphson method to find a root of the equation $x^2 \ln(x) = 5$, using $x_0 = 2$.

(d) The continuous random variable X has probability density function:

$$f(x) = \begin{cases} kx^3 & 1 \le x \le 4 \\ 0 & \text{otherwise.} \end{cases}$$

Determine the value of k and $P(1 < X < 2)$.

(e) The continuous random variable X has probability density function (PDF):

$$f(x) = \begin{cases} \frac{3}{8}x^2 & 0 \le x \le 2 \\ 0 & \text{otherwise.} \end{cases}$$

Find the associated cumulative distribution function (CDF) $F(x)$. Plot the PDF and CDF on the same graph.

(f) A report states that employees spend, on average, 80 minutes every working day on personal use of the Internet. A company takes a random sample of 100 employees and finds their mean personal Internet use is 83 minutes with a standard deviation of 15 minutes. The company's managing director claims that his employees spend more time on average on personal use of the Internet than the report states. Test, at the 5% level of significance, the managing director's claim.

5.4 Use Python to answer these questions.

(a) A stone is released from rest on a bridge and falls vertically into a lake. The stone has velocity $14\,\text{ms}^{-1}$ when it enters the lake. Calculate the distance the stone falls before it enters the lake, and the time after its release when it enters the lake.

(b) What is the resultant force of $F_1 = 2\underline{i} - 3\underline{j}$, $F_2 = -4\underline{i} + 5\underline{j}$ and $F_3 = 6\underline{i} + 3\underline{j}$. Find the magnitude of the resultant force.

(c) A 6m long uniform beam AB of weight 40N is supported at $A = 0$m by a vertical reaction R. The beam is held horizontal by a vertical wire attached at $C = 5$m from A. A particle of weight 30N is placed at 2m from A. Plot a figure representing the problem. Find the tension, T say, in the wire and the force R.

(d) A projectile is launched from the ground and must clear a wall of height 10 metres, a distance 100 metres away. Given that the initial velocity in the horizontal direction is $20\,\text{ms}^{-1}$, what is the minimum initial velocity required in the vertical direction for the projectile to clear the wall?

(e) A 2kg mass in limiting equilibrium rests on a rough plane inclined at an angle of 30 degrees to the horizontal. Show that the coefficient of friction between the mass and the plane is $\mu = \frac{1}{\sqrt{3}}$. If the plane is inclined at 40 degrees, how fast will the mass accelerate down the slope?

Solutions to the Exercises may be found here:

https://drstephenlynch.github.io/webpages/Solutions_Section_1.html.

FURTHER READING

[1] CGP Books. (2022). *New A-Level Maths AQA Complete Revision and Practice with Online Edition.* Coordination Group Publications Ltd.

[2] CGP Books. (2017). *A-Level Maths for AQA: Year 2 Student Book with Online Edition.* Coordination Group Publications Ltd.

II

Python for Scientific Computing

## CHAPTER 6

Biology

Biology is the branch of science that primarily deals with the structure, function, growth, evolution and distribution of organisms The first example is a simple discrete population model of a species of insect. Fixed points, stability and bifurcation diagrams are covered. The second example is a continuous population model of the interaction of predator-prey species. Critical points, stability and phase portraits are introduced. The third example is a continuous model of the spread of flu in a school. Compartmental models are used to model the spread of the virus and time series plots show how the populations vary. The final example is taken from a recent research paper and shows hysteresis in single fiber muscle.

Aims and Objectives:

- To use Python to study examples in biology.

At the end of the chapter, the reader should be able to

- find fixed points of period one and determine their stability;

- plot a bifurcation diagram;

- find critical points and determine their stability;

- plot phase portraits;

- use data in a model;

- interpret the solutions in real-world terms.

6.1   A SIMPLE POPULATION MODEL

In 1976, Robert May [3] showed that a simple nonlinear discrete dynamical equation was able to show complex, chaotic behavior. May's logistic map is written:

$$x_{n+1} = \mu x_n \left(1 - x_n\right), \tag{6.1}$$

where x_n represents the population of the single species, scaled from zero to one, by setting the parameter μ, $0 \le \mu \le 4$. The equation includes a simple reproduction

DOI: 10.1201/9781003285816-6

rate (the positive term) and a starvation rate (the negative term). Equation (6.1) is a discrete, nonlinear, one-dimensional equation. To find the population at time $t = n + 1$, simply insert the population at time, $t = n$. Physically, one can think of a large tank containing of a certain species of insect with a plentiful food supply and time is measured in days. The population is scaled between zero and one, an empty tank giving a population zero, and a full tank a population of one. For insects, the parameter μ could be used to measure the scaled temperature in the tank.

Example 6.1.1. Given $x_0 = 0.2$ and $t = 50$ days, use Python to investigate how the population varies when: (i) $\mu = 0.5$, (ii) $\mu = 1.5$, (iii) $\mu = 3.1$ and (iv) $\mu = 4$.

Solution. The Python program below is used for plotting a time series solution of the logistic map. Figure 6.1, shows the time series solutions. When $\mu = 0.5$, the temperature in the tank is too low to sustain life, so the insects do not reproduce, and the population decrease to zero. As the scaled temperature increases to $\mu = 1.5$, the population stabilises to a period-one solution, $x_n = x_S = \frac{1}{3}$. When $\mu = 3.1$, the population oscillates between $x_1 = 0.5580$ and $x_2 = 0.7646$, on a stable two-day cycle. Finally, when $\mu = 4$, the time series appears to be chaotic, or unpredictable.

```
# Program_6a.py: Iteration of the logistic map function.
import numpy as np
import matplotlib.pyplot as plt
mu , x = 4 , 0.2            # For case (iv).
xs = [0.2]                  # Initially, 20% of the tank is full.
for i in range(50):
    x = mu * x * (1 - x)
    xs = np.append(xs , x)
    #print(x)               # Print the x values if you like.
plt.plot(xs)
plt.xlabel("n")
plt.ylabel("$x_n$")
plt.show()
```

Example 6.1.2. Given $\mu = 4$ and (i) $x_0 = 2$, check for sensitivity to initial conditions by setting, (ii) $x_0 = 0.2001$.

Solution. Figure 6.2 shows the time series solutions when $x_0 = 0.2$ and $x_0 = 0.2001$. There is clearly sensitivity to initial conditions after about 10 days.

Definition 6.1.1. A mapping $x_{n+1} = f(x_n)$, has a fixed point of period one, x_S say, which satisfies the equation:

$$x_S = f(x_S). \tag{6.2}$$

The fixed point of period one is stable if and only if:

$$\left| \frac{df}{dx} \right|_{x_S} < 1. \tag{6.3}$$

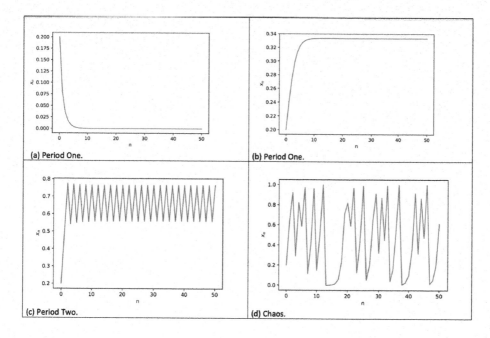

Figure 6.1 Time series for the logistic map function (6.1) when (a) $\mu = 0.5$, period one behavior; (b) $\mu = 1.5$, period-one behavior; (c) $\mu = 3.1$, period-two behavior and (d) $\mu = 4$, chaotic behavior.

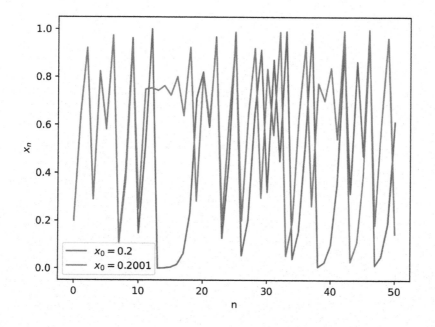

Figure 6.2 Time series for the logistic map function (6.1) when $\mu = 4$, and (i) $x_0 = 0.2$ and (ii) $x_0 = 0.2001$. The system is sensitive to initial conditions, which is one sign of chaos.

This is known as the stability condition. Solving equation (6.2) gives the fixed points of period one, and equation (6.3) is used to determine their stability.

Example 6.1.3. Use Python to plot a bifurcation diagram for the logistic map equation (6.1).

```
# Program_6b.py: A bifurcation diagram.
import numpy as np
import matplotlib.pyplot as plt
def f(x, mu):
    return mu * x * (1-x)
xs , mus = [] , np.linspace(0, 4, 40000)
for mu in mus:
  x = 0.1
  for i in range(500):
    x = f(x, mu)
  for i in range(50):
    x = f(x, mu)
    xs.append([mu, x])
xs = np.array(xs)
plt.plot(xs[: , 0] , xs[: , 1] , "r,")          # Pixel points.
plt.xlabel("$\mu$" , fontsize = 15)
plt.ylabel("x" , fontsize = 15)
plt.tick_params(labelsize = 15)
plt.show()
```

Solution. Figure 6.3 shows the bifurcation diagram, which summarises the behaviors of the dynamical system as the parameter μ varies, $0 \leq \mu \leq 4$. When $0 \leq \mu \leq 1$, there is a stable fixed point at $x_S = 0$. Physically, the temperature of the tank is too low to sustain life. As the scaled temperature increases between $1 \leq \mu \leq 3$, the system shows period-one, or steady-state behavior, and the population of insects slowly increases. When $\mu > 3$, there is a bifurcation to stable period-two behavior. There are then further bifurcations to period-4, period-8 etc. The chaotic region for approximately, $3.5 < \mu \leq 4$, is interspersed with periodic windows. Figure 6.3 has some fractal structure, readers can zoom in to the picture by editing Program_6b.py.

6.2 A PREDATOR-PREY MODEL

In 1975, Tanner published a paper on the intrinsic growth rates of predator and prey populations [5]. The predator-prey model was listed as a two-dimensional system of differential equations:

$$\frac{dx}{dt} = P(x,y) = rx\left(1 - \frac{x}{K}\right) - \frac{mx}{A+x}y, \quad \frac{dy}{dt} = Q(x,y) = sy\left(1 - h\frac{y}{x}\right), \quad (6.4)$$

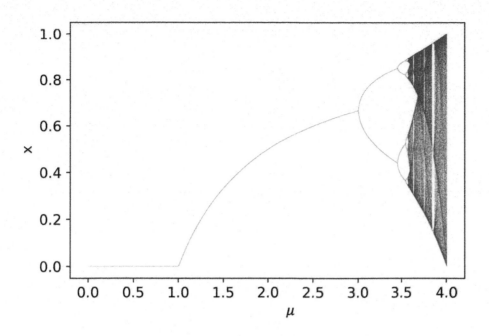

Figure 6.3 Bifurcation diagram for the logistic map function (6.1).

where $x(t)$ is the population of prey and $y(t)$ is the population of predator. The parameters m, K and A are constants, and the prey and predators grow logistically with intrinsic strengths, r and s, respectively, and h is the number of prey required to support one predator at equilibrium. This model is usually referred to as the Holling-Tanner model in the literature.

Definition 6.2.1. The critical points of system (6.4) are found by solving the simultaneous equations, $\dot{x} = 0$ and $\dot{y} = 0$, where $\dot{x} = \frac{dx}{dt}$. Denote a critical point by (x^*, y^*).

Definition 6.2.2. The Jacobian matrix of system (6.4) is defined by:

$$J = \begin{pmatrix} \frac{\partial P}{\partial x} & \frac{\partial P}{\partial y} \\ \frac{\partial Q}{\partial x} & \frac{\partial Q}{\partial y} \end{pmatrix},$$

where $\frac{\partial P}{\partial x}$ is the partial derivative of P with respect to x.

Definition 6.2.3. A critical point, say (x^*, y^*), of system (6.4) is stable if and only if all of the eigenvalues of the Jacobian matrix $J(x^*, y^*)$ have real part strictly negative. Otherwise, the critical point is unstable.

Example 6.2.1. Determine the critical points of system (6.4) and their stability, when $r = 1, K = 7, m = \frac{6}{7}, A = 1, s = 0.2$ and $h = 0.5$. Use Python to plot a time series and phase portrait of the solution curves, given that the initial scaled populations are $(x_0, y_0) = (7, 0.1)$. Give a physical interpretation of the result.

Solution. To find the critical points, set $\dot{x} = \dot{y} = 0$, and use the solve command in Python:

```
from sympy import *
x , y = symbols("x y")
solve([x*(1-x/7)-6*x*y/(7+7*x),0.2*y*(1-0.5*y/x)] , [x,y])
```

The solutions are $A = (1, 2)$, $B = (7, 0)$ and $C = (-7, -14)$. Straight away, we can discard the critical point C, as we can only have populations greater than or equal to zero.

The Jacobian matrix can be computed with Python using the program below.

```
# Program_6c.py: Determining the stability of critical points.
from sympy import symbols , diff , Matrix
x , y = symbols("x y")
P , Q = x*(1-x/7)-6*x*y/(7+7*x) , 0.2*y*(1-0.5*y/x)
J = Matrix([[diff(P,x),diff(P,y)],[diff(Q,x),diff(Q,y)]])
JA=J.subs([(x,1),(y,2)])
eigJA=JA.eigenvals()
print(eigJA)
JB=J.subs([(x,7),(y,0)])
eigJB=JB.eigenvals()
print(eigJB)
```

Therefore, the eigenvalues of the critical point at A are: $\lambda_1 = 0.0429 - 0.3353i$, with real part 0.0429, and $\lambda_2 = 0.0429 + 0.3353i$, with real part 0.0429, and the critical point A is unstable. The eigenvalues of the critical point at B are: $\lambda_1 = -1$, and $\lambda_2 = 0.2$, and the critical point B is unstable.

```
# Program_6d.py: Plot time series and a phase portrait.
import numpy as np
from scipy import integrate
import matplotlib.pyplot as plt
# The Holling-Tanner model.
def Holling_Tanner(X, t=0):
    # here X[0] = x and X[1] = y
    return np.array([ X[0]*(1-X[0]/7)-6*X[0]*X[1]/(7+7*X[0]), \
    0.2*X[1]*(1-0.5*X[1]/X[0]) ])
t = np.linspace(0, 200, 1000)
Sys0 = np.array([7, 0.1])               # Initial values: x0 = 7, y0 = 0.1.
X, infodict = integrate.odeint(Holling_Tanner,Sys0,t,full_output=True)
x,y = X.T
fig = plt.figure(figsize=(15,5))
fig.subplots_adjust(wspace = 0.5, hspace = 0.3)
```

```
ax1 = fig.add_subplot(1,2,1)
ax2 = fig.add_subplot(1,2,2)
ax1.plot(t,x, "r-",label="prey")
ax1.plot(t,y, "b-",label="predator")
ax1.set_title("Time Series")
ax1.set_xlabel("time")
ax1.grid()
ax1.legend(loc="best")
ax2.plot(x, y, color="blue")
ax2.set_xlabel("x")
ax2.set_ylabel("y")
ax2.set_title("Phase portrait")
ax2.grid()
plt.show()
```

Figure 6.4 shows the time series and phase portrait for the system (6.4). There is periodic behavior and the populations oscillate in phase with one another. The solution curve on the right in the phase portrait is an isolated periodic trajectory known as a limit cycle. When modeling two-dimensional systems in the real world, this is the most common form of solution. As an example of a physical model, this system can be applied to the populations of lynx (predator) and snowshoe-hare (prey) in Canada.

6.3 A SIMPLE EPIDEMIC MODEL

A simple model for COVID-19 is presented in [1], where the authors consider a susceptible, latently infected, symptomatic and asymptomatic infectious, and removed individuals (SLIAR) compartmental model. See the exercises at the end of the chapter for more details. In this section, we will consider a simpler SIR model described

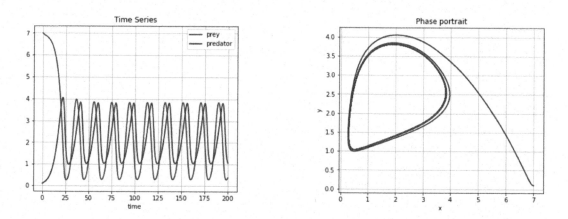

Figure 6.4 Time series and phase portrait for the predator-prey system (6.4).

by the differential equations:

$$\frac{dS}{dt} = -\frac{\beta SI}{N}, \quad \frac{dI}{dt} = \frac{\beta SI}{N} - \gamma I, \quad \frac{dR}{dt} = \gamma I, \tag{6.5}$$

where $S(t)$ is the susceptible population, $I(t)$ is the infected population, $R(t)$ is the recovered-immune population, N is the total population, β is the contact rate of the disease and γ is the mean recovery rate.

Exercise 6.3.1. Use Python to plot the solution curves for $S(t), I(t), R(t)$, given that $N = 1000$, $S(0) = 999$, $I(0) = 1$, $R(0) = 0$, $\beta = 0.5$ and $\gamma = 0.1$. This could be a simple model of the spread of flu in a school.

Solution. Program_6e.py is used to plot Figure 6.5.

```
# Program_6e.py: SIR Epidemic model.
import numpy as np
import matplotlib.pyplot as plt
from scipy.integrate import odeint
# Set the parameters.
beta, gamma = 0.5, 0.1
S0, I0, R0, N = 999, 1, 0, 1000
tmax, n = 100, 1000
def SIR_Model(X, t, beta, gamma):
    S, I, R = X
    dS = - beta * S * I / N
    dI = beta * S * I / N - gamma * I
    dR = gamma * I
    return(dS, dI, dR)
t = np.linspace(0, tmax, n)
f = odeint(SIR_Model, (S0, I0, R0), t, args = (beta, gamma))
S, I, R = f.T
plt.figure(1)
plt.xlabel("Time (days)")
plt.ylabel("Populations")
plt.title("Susceptible-Infected-Recovered (SIR) Epidemic Model")
plt.plot(t, S, label = "S")
plt.plot(t, I, label = "I")
plt.plot(t, R, label = "R")
legend = plt.legend(loc = "best")
plt.show()
```

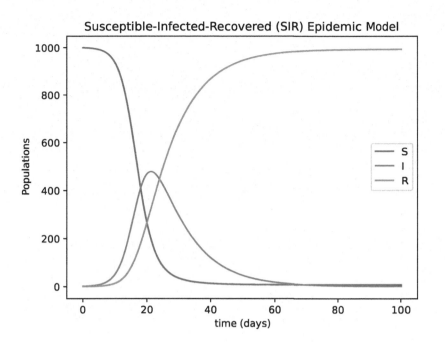

Figure 6.5 Time series solutions for the Susceptible-Infected-Recovered (SIR) epidemic model (6.5).

6.4   HYSTERESIS IN SINGLE FIBER MUSCLE

In 2017, Ramos et al. investigated mice single fiber muscle under lengthening, holding, and shortening [4]. It was shown that there was hysteresis present in the system.

Definition 6.4.1. Hysteresis is the dependence of the state of a system on its history.

Example 6.4.1. Reproduce the results from the paper [4] to obtain Figure 20 from that paper.

Solution. The program to create the data for the fraction of muscle length against time, see Figure 6.6, is listed below. The program also solves Hill's ODEs giving Figure 6.7.

```
# Program_6f.py: Hysteresis in Single Fiber Muscle.
# Create data for muscle model.
import numpy as np
import matplotlib.pyplot as plt
Length , a , b = 1200 , 380*0.098 , 0.325
F0 = a / 0.257          # Initial force.
vm , alpha , LSE0 , k = F0*b/a , F0/0.1 , 0.3 , a/25
```

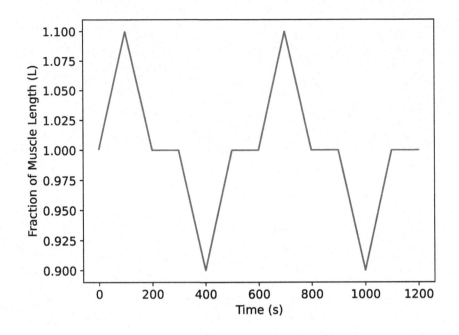

Figure 6.6 Data generated with Python. Time against fraction of muscle length (L).

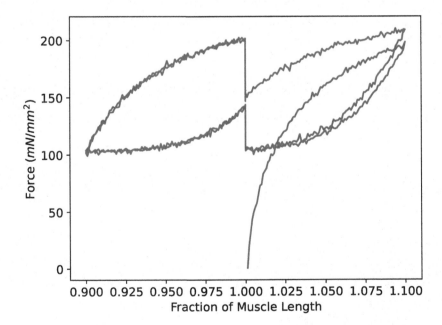

Figure 6.7 A graph of fraction of muscle length against force. The solution curve is showing hysteresis.

```
t = [0+0.01*i for i in range(1201)]        # Time.
A = [1.001+0.001*i for i in range(100)]
B = [1.099-0.001*i for i in range(100)]
C = np.ones(100).tolist()
D = [0.999-0.001*i for i in range(100)]
E = [0.901+0.001*i for i in range(100)]
F = np.ones(100).tolist()
G = [1.001+0.001*i for i in range(100)]
H = [1.099-0.001*i for i in range(100)]
HH = np.ones(100).tolist()
J = [0.999-0.001*i for i in range(100)]
K = [0.901+0.001*i for i in range(100)]
KK = np.ones(101).tolist()
L = A+B+C+D+E+F+G+H+HH+J+K+KK
plt.figure()
plt.plot(L)
plt.xlabel("Time (s)", fontsize=12)
plt.ylabel("Fraction of Muscle Length (L)", fontsize=12)
plt.tick_params(labelsize=12)
plt.show()
# Solve Hill's ODEs."
LSE = np.zeros(1200).tolist() # Length series element.
LCE = np.zeros(1200).tolist() # Length contractile element.
F = np.zeros(1201).tolist()
for i in range(1200):          # Hill's ODEs.
    LSE[i] = 0.3 + F[i]/alpha
    LCE[i] = L[i] - LSE[i]
    dt = t[i+1] - t[i]
    dL = L[i+1] - L[i]
    dF = alpha*((dL/dt)+b*((F0-F[i])/(a+F[i])))*dt
    F[i+1] = F[i] + dF
F = np.array(F)
FF = (F0/100)*np.random.randn(1201)
F = F + FF
F = F.tolist()
plt.figure()
plt.plot(L,F)
plt.xlabel("Fraction of Muscle Length", fontsize=12)
plt.ylabel("Force ($mN/mm^2$)", fontsize=12)
plt.tick_params(labelsize=12)
plt.show()
```

Figure 6.6 shows that the single fiber of muscle is lengthened (eccentric contraction) for $0 \leq t \leq 200$ seconds, then held (isometric contraction) for $200 \leq t \leq 300$ seconds, shortened (concentric contraction) for $300 \leq t \leq 500$ seconds, and then held for $500 \leq t \leq 600$ seconds. The cycle is repeated for $600 \leq t \leq 1200$ seconds.

Figure 6.7 shows that the single fiber of muscle undergoes hysteresis as it is lengthened, held and shortened on a periodic cycle. The results from the mathematical model match very well with the results from experiment.

EXERCISES

6.1 Blood cells in the body are continually being created and destroyed. Consider the blood-cell population model described by:

$$c_{n+1} = (1 - a)c_n + bc_n^r e^{-sc_n},$$

where c_n is the blood cell count, $0 < a \leq 1$, and $b, r, s > 0$. Assuming that $b = 1.1 \times 10^6$, $r = 8$ and $s = 16$, show that there are (i) three fixed points, two stable and one unstable, when $a = 0.2$, and (ii) three fixed points, two unstable and one stable, when $a = 0.3$. For clinical examples, read the author's paper [2] on the analysis of a blood cell population model.

6.2 A predator-prey system is modeled using the differential equations:

$$\dot{x} = 4x - x^2 - xy, \quad \dot{y} = -y - y^2 + 3xy,$$

where $x(t)$ is the population of prey and $y(t)$ is the population of predator. Determine the number and location of the critical points and plot a phase portrait with a number of solution curves of your choice. How would you describe the phase portrait in physical terms?

6.3 A simple SLIAR model of the spread of coronavirus in Manchester can be depicted by the compartmental model shown in Figure 6.8. For this model, β is the

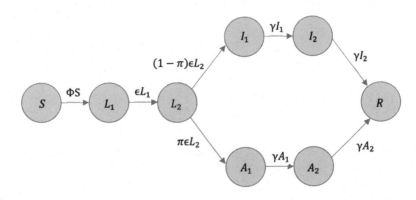

Figure 6.8 Compartmental model of the spread of coronavirus in Manchester. In this case $\Phi = \beta \left(I_1 + I_2 + \xi \left(A_1 + A_2 \right) + \eta L_2 \right)$, is the force of infection, S is the susceptible population, L_1, L_2 are latently infectious, I_1, I_2 are infectious, A_1, A_2 are asymptomatic, and R is the removed population.

transmission coefficient, η and ξ, are the attenuation factors for transmission by incubating and asymptomatic cases, respectively, π denotes the split between infectious and asymptomatic infectious individuals, and ϵ and γ, describe the rates at which incubation and infectiousness end, respectively.

(a) Write down the differential equations that describe the model (see [1]).

(b) Use Python, to plot time series solution curves given, $\beta = 10^{-4}$, $\eta = 0.2$, $\xi = 0.1$, $\pi = 0.1$ $\epsilon = 0.01$, $\gamma = 0.15$, $S(0) = 2,700,000$, $I_1(0) = 1$, $L_1(0) = L_2(0) = A_1(0) = A_2(0) = I_2(0) = R(0) = 0$ and $0 \leq t \leq 1000$ days.

(c) Determine the maximum values of I_1 and A_1.

6.4 Edit Programs 6f and 6g to show what happens if the single fiber muscle is simply lengthened and shortened without holding. Plot the fraction of muscle length against force.

Solutions to the Exercises may be found here:

https://drstephenlynch.github.io/webpages/Solutions_Section_2.html.

FURTHER READING

[1] Arino, J. and Portet, S. (2020). A simple model for COVID-19. *Infect. Dis. Model.* **5**, 309–315.

[2] Lynch, S. (2005). Analysis of a blood cell population model. *Int. J. of Bifurcation and Chaos* **15**, 2311–2316.

[3] May, R. (1976). Simple mathematical models with very complicated dynamics. *Nature* **261**(5560), 459–467.

[4] Ramos, J., Lynch, S., Jones, D.A. and Degens, H. (2017). Hysteresis in muscle (Feature article). *Int. J. of Bifurcation and Chaos* **27**(5560) 1730003, 1–16.

[5] Tanner, J.T. (1975). The stability and the intrinsic growth rates of prey and predator populations. *Ecology* **56**, 855–867.

## CHAPTER 7

Chemistry

Chemistry is the branch of science that deals with the properties, composition, and structure of elements and compounds, how they can change, and the energy that is released or absorbed when they change. The first example uses matrices and null-space vectors to determine the coefficients of balanced chemical-reaction equations. The second and third examples are taken from chemical kinetics, using differential equations to determine the amounts of reactants and products after a given time. The final example covers the common-ion effect in solubility for silver chloride in a potassium chloride solution.

Aims and Objectives:

- To use Python to study examples in chemistry.

At the end of the chapter, the reader should be able to

- balance chemical-reaction equations;

- use numerical methods to solve problems in chemical kinetics;

- plot solubility curves for common ions;

- interpret the solutions in real-world terms.

7.1 BALANCING CHEMICAL-REACTION EQUATIONS

In 2010, Thorne [8] demonstrated an innovative approach using the matrix null-space method, that is applicable to all chemical-reaction equations. We will illustrate the method with the same simple example used by Thorne, and then show how the problem can be solved with the aid of Python. Consider the following chemical-reaction equation:

$$x_1 \text{KI} + x_2 \text{KClO}_3 + x_3 \text{HCl} \rightleftharpoons x_4 \text{I}_2 + x_5 \text{H}_2\text{O} + x_6 \text{KCl}, \tag{7.1}$$

where x_1 to x_6, are the coefficients to be found, with elemental constituents, K is potassium, I is iodine, O is oxygen, H is hydrogen and Cl is chlorine.

DOI: 10.1201/9781003285816-7

Example 7.1.1. Use the matrix null-space method to determine the coefficients x_1 to x_6, for equation (7.1).

Solution. First of all, we construct a chemical composition table for equation (7.1), indicating the numbers of atoms of each of the chemical elements which make up the reactants (on the left hand side of the equation) and the products (on the right-hand side of the equation).

Table 7.1   Chemical Composition Table

Element	KI	KClO$_3$	HCl	I$_2$	H$_2$O	KCl
K	1	1	0	0	0	1
I	1	0	0	2	0	0
O	0	3	0	0	1	0
H	0	0	1	0	2	0
Cl	0	1	1	0	0	1

We can write the numbers of the chemical-composition table (see Table 7.1) in matrix form. The chemical-composition matrix, CCM say, is written as:

$$CCM = \begin{pmatrix} 1 & 1 & 0 & 0 & 0 & 1 \\ 1 & 0 & 0 & 2 & 0 & 0 \\ 0 & 3 & 0 & 0 & 1 & 0 \\ 0 & 0 & 1 & 0 & 2 & 0 \\ 0 & 1 & 1 & 0 & 0 & 1 \end{pmatrix}.$$

In order to make the matrix square, we construct an augmented chemical-composition matrix, $ACCM$ say, as illustrated below:

$$ACCM = \left(\begin{array}{cccccc} 1 & 1 & 0 & 0 & 0 & 1 \\ 1 & 0 & 0 & 2 & 0 & 0 \\ 0 & 3 & 0 & 0 & 1 & 0 \\ 0 & 0 & 1 & 0 & 2 & 0 \\ 0 & 1 & 1 & 0 & 0 & 1 \\ \hline 0 & 0 & 0 & 0 & 0 & 1 \end{array} \right).$$

Using Python, we can find the inverse of this matrix and then use the null-space vector (from the last column) to determine the coefficients x_1 to x_6, as shown in Program_7a.py. Note that each vector element is scaled (or divided) by the element with the smallest absolute value so that the smallest number in the null-space vector is one. The negative and positive values correspond to coefficients on the left and right of the chemical-reaction equation, respectively.

```
# Program_7a.py: Compute the matrix null-space vector.
from sympy import Matrix
# Construct the augmented matrix.
```

```
ACCM=Matrix([[1,1,0,0,0,1],\
        [1,0,0,2,0,0],\
        [0,3,0,0,1,0],\
        [0,0,1,0,2,0],\
        [0,1,1,0,0,1],\
        [0,0,0,0,0,1]])
print(ACCM)
invACCM=ACCM.inv() # Find the inverse matrix.
print(invACCM)
Nullv=invACCM.col(5) / min(abs(invACCM.col(5))) # Last column.
print(Nullv) # Scaled null-space vector.
```

The solution is, Nullv=Matrix([[−6], [−1], [−6], [3], [3], [7]]), giving

$$x_1 = 6, x_2 = 1, x_3 = 6, x_4 = 3, x_5 = 3, x_6 = 7,$$

and the balanced chemical-reaction equation is

$$6KI + KClO_3 + 6HCl \rightleftharpoons 3I_2 + 3H_2O + 7KCl.$$

7.2   CHEMICAL KINETICS

Chemical kinetics is a branch of physical chemistry concerned with understanding rates of reaction. Even the simplest chemical reactions can be highly complex and difficult to model, factors such as physical state, surface area, concentrations, temperature, pressure, absorption of light and the use of catalysts, for example, can all affect the rates. Unfortunately, we cannot write down the differential equations for the chemical reaction based on the reaction rate equations alone. For more information, the reader is directed to the books [5] and [6]. In this chapter, we will only be concerned with the differential equations used to model the chemical reaction.

Example 7.2.1. The chemical equation for the reaction between nitrous oxide and oxygen to form nitrogen dioxide at 25^oC,

$$2NO + O_2 \rightleftharpoons 2NO_2, \tag{7.2}$$

obeys the law of mass action. The rate equation is given by

$$\frac{dc}{dt} = k \left(a_0 - c\right)^2 \left(b_0 - \frac{c}{2}\right),$$

where $c = [NO_2]$ is the concentration of nitrogen dioxide, k is the rate constant, a_0 is the initial concentration of NO and b_0 is the initial concentration of O_2. Find the concentration of nitrogen dioxide after time $t = 100$ seconds, given that $k = 0.00713\, l^2M^{-2}s^{-1}$, $a_0 = 4Ml^{-1}$, $b_0 = 1Ml^{-1}$ and $c_0 = 0Ml^{-1}$. Plot the solution curve.

Solution. The program for finding the concentration after 100 seconds, and plotting the solution curve is listed below. We solve the differential equation numerically using the **odeint** solver.

```python
# Program_7b.py: Production of Nitrogen Dioxide.
import numpy as np
import matplotlib.pyplot as plt
from scipy.integrate import odeint
k , a0 , b0 , c0 = 0.00713 , 4 , 1, 0
def ode(c , t):
    dcdt = k * (a0 - c)**2 * (b0 - c / 2)
    return dcdt
t = np.linspace(0 , 400 , 401) # t=[0,1,2,...,400]
c = odeint(ode , c0 , t)
plt.xlabel("Time (s)")
plt.ylabel("c(t) (Ml$^{-1})$")
plt.title("Production of Nitrogen Dioxide")
plt.plot(t , c)
print("c(100) = ", c[100], "Moles per litre")
plt.show()
```

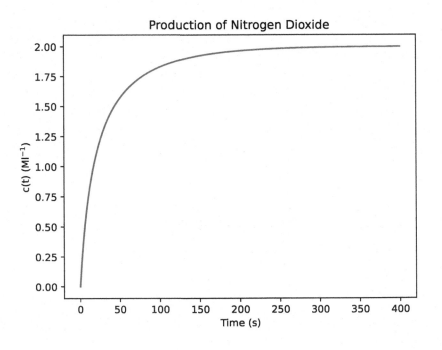

Figure 7.1 The solution curve for the ODE chemical-reaction equation (7.2). The program gives $c(100) = 1.828819$ Moles per litre.

7.3 THE BELOUSOV-ZHABOTINSKI REACTION

Oscillating chemical reactions such as the Bray-Liebhafsky reaction [2], the Briggs-Rauscher reaction [3] and the Belousov-Zhabotinski (BZ) reaction [7] provide wonderful examples of relaxation oscillations in science. The first experiment of the BZ reaction was conducted by Boris Belousov in the 1950s, and the results were not confirmed until as late as 1968 by Zhabotinski. Following the methods of Field and Noyes [4], the chemical rate equations for an oscillating BZ reaction are written as:

$$
\begin{aligned}
&\text{BrO}_3^- + \text{Br}^- \rightarrow \text{HBrO}_2 + \text{HOBr}, &&\text{Rate} = k_1[\text{BrO}_3^-][\text{Br}^-] \\
&\text{HBrO}_2 + \text{Br}^- \rightarrow 2\text{HOBr}, &&\text{Rate} = k_2[\text{HBrO}_2][\text{Br}^-] \\
&\text{BrO}_3^- + \text{HBrO}_2 \rightarrow 2\text{HBrO}_2 + 2\text{M}_{\text{OX}}, &&\text{Rate} = k_3[\text{BrO}_3^-][\text{HBrO}_2] \\
&2\text{HBrO}_2 \rightarrow \text{BrO}_3^- + \text{HOBr}, &&\text{Rate} = k_4[\text{HBrO}_2]^2 \\
&\text{OS} + \text{M}_{\text{OX}} \rightarrow \tfrac{1}{2}C\text{Br}^-, &&\text{Rate} = k_5[\text{OS}][\text{M}_{\text{OX}}]
\end{aligned}
$$

where OS represents all oxidizable organic species and C is a constant. Note that in the third equation, species HBrO_2 stimulates its own production, a process called *autocatalysis*. The reaction rate equations for the concentrations of intermediate species $x = [\text{HBrO}_2]$, $y = [\text{Br}^-]$ and $z = [\text{M}_{\text{OX}}]$ are

$$
\begin{aligned}
\frac{dx}{dt} &= k_1 a y - k_2 x y + k_3 a x - 2k_4 x^2, \\
\frac{dy}{dt} &= -k_1 a y - k_2 x y + \frac{1}{2}C k_5 b z, \\
\frac{dz}{dt} &= 2k_3 a x - k_5 b z,
\end{aligned}
\tag{7.3}
$$

where $a = [\text{BrO}_3^-]$ and $b = [\text{OS}]$ are assumed to be constant, and $[\text{M}_{\text{OX}}]$ represents the metal ion catalyst in its oxidized form. Taking the transformations

$$
X = \frac{2k_4 x}{k_5 a}, \quad Y = \frac{k_2 y}{k_3 a}, \quad Z = \frac{k_5 k_4 b z}{(k_3 a)^2}, \quad \tau = k_5 b t,
$$

system (7.3) becomes

$$
\begin{aligned}
\frac{dX}{d\tau} &= \frac{qY - XY + X(1 - X)}{\epsilon_1}, \\
\frac{dY}{d\tau} &= \frac{-qY - XY + CZ}{\epsilon_2}, \\
\frac{dZ}{d\tau} &= X - Z,
\end{aligned}
\tag{7.4}
$$

where X, Y, Z are the concentrations of bromous acid, bromide ions and cerium ions, respectively, $\epsilon_1 = \frac{k_5 b}{k_3 a}$, $\epsilon_2 = \frac{2k_5 k_4 b}{k_2 k_3 a}$, and $q = \frac{2k_1 k_4}{k_2 k_3}$.

Example 7.3.1. Using the dimensionless system (7.4), plot the relative concentrations X, Y, Z for $0 \leq t \leq 50$, given that $X(0) = 0$, $Y(0) = 0$, $Z(0) = 0.1$, $\epsilon_1 = 0.0099$, $\epsilon_2 = 2.4802 \times 10^{-5}$, $q = 3.1746 \times 10^{-5}$ and $C = 1$.

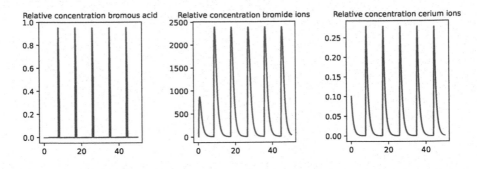

Figure 7.2 Periodic behavior for the stiff BZ ODE (7.4). Physically, a clear solution goes dark pink, and then over the next 10 seconds goes clearer and clearer, and then suddenly goes dark pink again. This process continues in an oscillatory fashion. It is often shown at university open days in chemistry laboratories.

Solution. This is a stiff system of ODEs (the system includes large and small parameters) and is intractable to analytical approaches. The only way to solve this system is with numerical methods using packages such as Python. The Python program is listed below and the oscillating solutions are displayed in Figure 7.2.

```
# Program_7c.py: The Belousov-Zhabotinski Reaction.
import numpy as np
from scipy.integrate import odeint
import matplotlib.pyplot as plt
# B_Z parameters and initial conditions.
q, f, eps1, eps2 = 3.1746e-5, 1, 0.0099, 2.4802e-5
x0, y0, z0 = 0, 0, 0.1
# Maximum time point and total number of time points.
tmax, n = 50, 10000
def B_ZReaction(X,t,q,f,eps1,eps2):
    x, y, z = X
    dx = (q*y - x*y + x*(1 - x))/eps1
    dy = ( - q*y - x*y + f*z)/eps2
    dz = x - z
    return dx, dy, dz
t = np.linspace(0, tmax, n)
f = odeint(B_ZReaction, (x0, y0, z0), t, args=((q,f,eps1,eps2)))
x, y, z = f.T
# Plot time series.
fig = plt.figure(figsize=(10,3))
fig.subplots_adjust(wspace = 0.5, hspace = 0.3)
ax1 = fig.add_subplot(1,3,1)
ax1.set_title("Relative concentration bromous acid", fontsize=10)
ax2 = fig.add_subplot(1,3,2)
ax2.set_title("Relative concentration bromide ions", fontsize=10)
```

```
ax3 = fig.add_subplot(1,3,3)
ax3.set_title("Relative concentration cerium ions", fontsize=10)
ax1.plot(t, x, "b-")
ax2.plot(t, y, "r-")
ax3.plot(t, z, "m-")
plt.show()
```

7.4   COMMON-ION EFFECT IN SOLUBILITY

Silver chloride, AgCl, is an important photosensitive inorganic material used in photography, it is also used in antidotes, antimicrobials, personal deodorants, wound healing materials and water treatment, for example. In nature, it is found as the mineral chlorargyrite, however, it is easily created in the laboratory using precipitation and the common ion effect. Silver chloride is a white crystalline solid with a low solubility. It is weakly ionized in its aqueous solution, and there exists an equilibrium:

$$AgCl(s) \rightleftharpoons Ag^+(aq) + Cl^-(aq),$$

where (s) represents a solid state and (aq) is the aqueous state. The solubility product equilibrium constant, K_{SP} for the reaction is:

$$K_{SP} = [Ag^+] \times [Cl^-]. \tag{7.5}$$

If this is the only reaction to consider, then we predict that the precipitate's solubility, S_{AgCl}, is given by:

$$S_{AgCl} = [Ag^+] = \frac{K_{SP}}{[Cl^-]}. \tag{7.6}$$

Definition 7.4.1. Le Chatelier's principle states, if a stress is applied to a reaction mixture at equilibrium, the net reaction goes in the direction that relieves the stress.

Example 7.4.1. Determine the molar solubility, x say, of AgCl in 0.1 M, potassium chloride, KCL(aq).

Solution. KCl is a very soluble salt and will dissolve completely into K^+(aq) and Cl^-(aq), and the concentration of each ion will be 0.1 M. We use an Initial-Change-Equilibrium (ICE) table, see Table 7.2.

Table 7.2   Dissociation of AgCl in 0.1M KCl

	$[Ag^+]$	$[Cl^-]$
Initial (I)	0	0.1
Change (C)	+x	+x
Equilibrium (E)	x	0.1 + x

Now K_{SP} for AgCl is a constant 1.82×10^{-10}, therefore, at equilibrium:

$$K_{SP} = [Ag^+][Cl^-] = x(0.1 + x),$$

and one solution of this quadratic equation is

$$x = 1.82 \times 10^{-9}M.$$

The molar solubility of AgCl in 0.1M KCl(aq), is much lower than the solubility in pure water $(1.33 \times 10^{-5}$ M) as predicted by Le Chatelier's principle.

There are three simultaneous equations which follow from these equations:

First equation: $[Ag^+] \times [Cl^-] = K_{SP}$.
Second equation: $[Ag^+] + [K^+] = [Cl^-]$.
Third equation: $[K^+] = [K^+]$.

These equations can be solved using the **fsolve** function in Python.

Example 7.4.2. Suppose that the silver chloride is immersed in a solution of potassium chloride. Use Python to plot a $\log(S_{AgCl})$ against concentration of potassium chloride graph for $0 \le |KCl| \le 0.1$ Moles per litre.

Solution. The program for plotting Figure 7.3 is listed below.

```
# Program_7d.py: Common Ion Effect Silver Chloride and Potassium
Chloride.
import numpy as np
import matplotlib.pyplot as plt
import scipy.optimize as opt
numpoints = 100
Ag_0 , AgCl2_0 , Cl_0 , K_0  = 0 , 0 , 0.1 , 0.1
initparams = (Ag_0 , Cl_0 , K_0)
def AgCl_sol(concentrations):
    (Ag, Cl, K) = concentrations
    firstEq = Ag * Cl - 1.82E-10
    secondEq = Ag + K - Cl
    thirdEq = K - K
    return[firstEq, secondEq, thirdEq]
solution = opt.fsolve(AgCl_sol,initparams)
solubility = "{:.2E}".format(solution[0])
print("When |KCl| is", K_0, "M, AgCl solubility is", solubility)
logxrange = np.logspace(-5, -1 , num = numpoints)
#Below are the starting points, with Ag+ = 0, K+ and Cl- = x-range
guess_array = tuple(zip(np.zeros(numpoints),logxrange,logxrange))
```

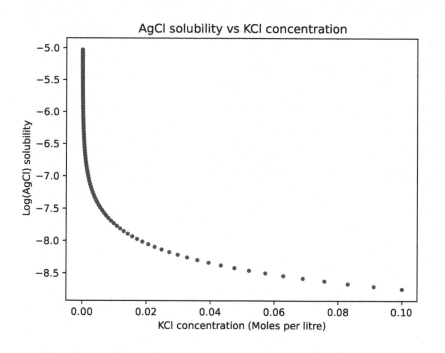

Figure 7.3 Solubility of AgCl as a function of KCl concentration.

```
out_array , Ag_conc , Cl_conc , K_conc = [] , [] ,[] ,[]
for num in range(0 , numpoints):
    out_array.append(list(opt.fsolve(AgCl_sol,guess_array[num])))
    Ag_conc.append(out_array[num][0])
    Cl_conc.append(out_array[num][1])
    K_conc.append(out_array[num][2])
plt.plot(K_conc,np.log10(Ag_conc),"r.")
plt.title("AgCl solubility vs KCl concentration")
plt.xlabel("KCl concentration (Moles per litre)")
plt.ylabel("Log(AgCl) solubility")
plt.show()
```

Equation (7.6) suggests that one can minimize solubility losses by adding a large excess of Cl⁻ ions. In fact, adding a large excess of chloride ions actually increases the precipitate's solubility. The reason for this is that silver ions also form a series of soluble silver-chlorometal-ligand complexes. By adding one more equation, see (7.7), a true graph of the solubility curve starts to form as shown in Figure 7.4.

$$Ag^+(aq) + 2Cl^-(aq) \rightleftharpoons AgCl_2^-(aq), \tag{7.7}$$

with $K_F = 1.78 \times 10^{-5}$. Readers will be asked to show this in the exercises in the next section. For a recent review article on the common-ion effect and its applications in chemistry, readers are directed to [1].

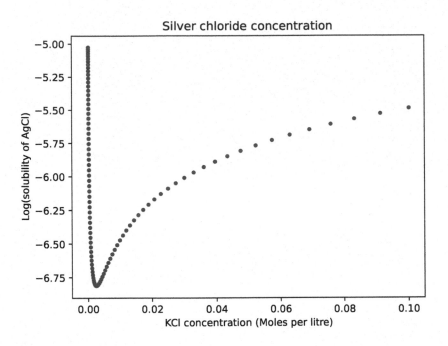

Figure 7.4 The true graph of the solubility of AgCl as a function of KCl concentration.

EXERCISES

7.1 Use the matrix null-space method to determine the coefficients x_1 to x_5, for the chemical-reaction equation:

$$x_1 NaHCO_3 + x_2 H_3 C_6 H_5 O_7 \rightleftharpoons x_3 Na_3 C_6 H_5 O_7 + x_4 H_2 O + x_5 CO_2.$$

Name the chemical elements involved.

7.2 The ODEs which model a catalytic reaction:

$$\left. \begin{array}{c} A + X \xrightarrow{k_1} R + Y \\ B + Y \xrightarrow{k_2} S + X \end{array} \right\} \Rightarrow A + B \rightarrow R + S,$$

are given by

$$\frac{da}{dt} = -k_1 ax, \quad \frac{db}{dt} = -k_2 by, \quad \frac{dx}{dt} = -k_1 ax + k_2 by, \quad \frac{dy}{dt} = k_1 ax - k_2 by,$$

where A and B are reactants, and X and Y are catalysts, with concentrations a, b, x and y, respectively. The products are R and S. Given that $k_1 = 0.1$, $k_2 = 0.05$, $a_0 = b_0 = x_0 = 1$ and $y_0 = 0$, plot the solution curves for $0 \leq t \leq 250$. What can you say about the long-term concentrations x and y?

7.3 The chemical rate equations for the Chapman cycle modeling the production of ozone are

$$O_2 + hv \rightarrow O + O, \qquad \text{Rate} = k_1,$$
$$O_2 + O + M \rightarrow O_3 + M, \quad \text{Rate} = k_2,$$
$$O_3 + hv \rightarrow O_2 + O, \qquad \text{Rate} = k_3,$$
$$O + O_3 \rightarrow O_2 + O_2, \qquad \text{Rate} = k_4,$$

where O is a singlet, O_2 is oxygen and O_3 is ozone. The reaction rate equations for species $x = [O]$, $y = [O_2]$ and $z = [O_3]$ are

$$\dot{x} = 2k_1 y + k_3 z - k_2 xy[M] - k_4 xz,$$
$$\dot{y} = k_3 z + 2k_4 xz - k_1 y - k_2 xy[M],$$
$$\dot{z} = k_2 xy[M] - k_3 z - k_4 xz.$$

This is another stiff system of ODEs. Given that $[M] = 9e17$, $k_1 = 3e{-}12$, $k_2 = 1.22e{-}33$, $k_3 = 5.5e{-}4$, $k_4 = 6.86e{-}16$ $x(0) = 4e16$, $y(0) = 2e16$ and $z(0) = 2e16$, show that the steady-state reached is $[O] = 4.6806e7, [O_2] = 6.999e16$ and $[O_3] = 6.5395e12$.

7.4 Use Python to reproduce Figure 7.4 by adding equation (7.7) into the mix. Readers should note that there are also other chemical ion equations which need to be considered.

Solutions to the Exercises may be found here:

https://drstephenlynch.github.io/webpages/Solutions_Section_2.html.

FURTHER READING

[1] Ayugo, P.C., Ezugwu, M.I. and Eze, F.I. (2020). Principle of common-ion effect and its applications in chemistry. *J. Chem. Lett.* **1**(3), 77-83.

[2] Bray, W.C. and Liebhafsky, H.A. (1931). Reactions involving hydrogen peroxide and iodate ion. *J. of the Amer. Chem. Soc.* **53**, 38-44.

[3] Briggs, T.S. and Rauscher W.C. (1976). An oscillating iodine clock. *J. Chem. Ed.* **50**, 496.

[4] Field, R. and Burger, M. (1985). *Oscillations and Travelling Waves in Chemical Systems*. Wiley, New York.

[5] Houston, P.L. (2006). *Chemical Kinetics and Reaction Dynamics*. Independently published.

[6] Patil, M.K. and Patil, S.S.M. (2020). *Chemical Kinetics and Photochemistry*. Dover Publications, New York.

[7] Scott, S.K. (1994). *Oscillations, Waves, and Chaos in Chemical Kinetics.* Oxford University Press, Oxford.

[8] Thorne, L.R. (2010). An innovative approach to balancing chemical-reaction equations: A simplified matrix-inversion technique to determine the matrix null space. *Chem. Educator* **15**, 304-308.

## CHAPTER 8

Data Science

Data Science is the field of study combining programming, statistics, mathematics and domain expertise to work with data. There are many excellent Python books covering the subject, see for example, [5], [7], [9] and [12]. Note that Data Science also encompasses machine learning and deep learning, which are covered in the last part of the book. The reader is first introduced to the Pandas library for working with data frames, then optimization problems are introduced for linear programming, and in the third section, k-means clustering is covered as a form of unsupervised learning. The final section covers classification decision trees for making predictions on data, and there is an example of a regression decision tree in the exercises at the end of the chapter.

Aims and Objectives:

- To display and work with data using data frames in Pandas.

- To introduce linear programming including the simplex method.

- To cluster data using unsupervised learning.

- To introduce decision trees.

At the end of the chapter, the reader should be able to

- Use Pandas to create data frames to be used in data science.

- Plot feasibility regions in linear programming and solve optimization problems.

- Use k-means clustering to order data.

- Create decision trees to make predictions.

8.1 INTRODUCTION TO PANDAS

Pandas (Python and Data Analysis) is an open source Python package used to solve a wide range of problems in data science. For more detailed information, the reader is directed to the URL:

DOI: 10.1201/9781003285816-8

https://pandas.pydata.org.

To illustrate the many uses of Pandas consider the following example.

Example 8.1.1. A pharmaceutical company is assessing the efficacy of a weight loss drug using 20 volunteers over the period of a month. The drugs are administered to 10 volunteers on one day, and another 10 volunteers the following day. Figure 8.1 shows the data frame, column 1 gives the date that the drug was administered, column 2, shows the dosage of the drug in milligrams column 3, lists the sex of the volunteers, column 4, lists their weight in kilograms and the final column shows the drugs efficacy, where the target is to lose one kilogram in a month. Use Pandas to create this data frame.

Solution. Figure 8.1 shows the data frame produced by running Program_8a,py listed below. Note that there are two data frames which are concatenated. The program shows how to create a dictionary of objects that can be converted into series-like format. A dictionary consists of key-value pairs, for example, in row zero, the key "Date" (2022-01-10) has a value pair "Dosage (mg)." The columns of the data frame have different data types, manipulation of the data is illustrated in the examples below.

```python
# Program_8a.py: Create a data frame for drug efficacy.
import numpy as np
import pandas as pd
df1=pd.DataFrame({
    "Date": pd.Timestamp("20220125"),
    "Dosage (mg)": np.array(list(range(2,22,2))),
    "Sex " : pd.Categorical(["F","F","M","F","M",
                              "M","F","F","F","M"]),
    "Weight (Kg)" : pd.Series([52.2,65.8,80.7,53.5,40.9,
                    52.2,64.4,61.7,53.5,61.2],
                    dtype="float32"),
    "Efficacy (%)" : pd.Series([0,0,0,0,5,
                        20,90,100,100,100],dtype="int32")
    })
df2=pd.DataFrame({
    "Date": pd.Timestamp("20220126"),
    "Dosage (mg)": np.array(list(range(22,42,2))),
    "Sex " : pd.Categorical(["M","M","F","M","M",
                              "F","M","F","F","M"]),
    "Weight (Kg)" : pd.Series([
                    86.2,72.6,59.0,67.1,56.2,
                    61.2,78.0,45.3,54.4,88.9],
                    dtype="float32"),
    "Efficacy (%)" : pd.Series([
```

	Date	Dosage (mg)	Sex	Weight (Kg)	Efficacy (%)
0	2022-01-25	2	F	52.200001	0
1	2022-01-25	4	F	65.800003	0
2	2022-01-25	6	M	80.699997	0
3	2022-01-25	8	F	53.500000	0
4	2022-01-25	10	M	40.900002	5
5	2022-01-25	12	M	52.200001	20
6	2022-01-25	14	F	64.400002	90
7	2022-01-25	16	F	61.700001	100
8	2022-01-25	18	F	53.500000	100
9	2022-01-25	20	M	61.200001	100
10	2022-01-26	22	M	86.199997	70
11	2022-01-26	24	M	72.599998	65
12	2022-01-26	26	F	59.000000	55
13	2022-01-26	28	M	67.099998	50
14	2022-01-26	30	M	56.200001	50
15	2022-01-26	32	F	61.200001	45
16	2022-01-26	34	M	78.000000	10
17	2022-01-26	36	F	45.299999	0
18	2022-01-26	38	F	54.400002	0
19	2022-01-26	40	M	88.900002	0

Figure 8.1 Pandas data frame for drug efficacy. The data frame was produced using Program_8a.py.

```
                        70,65,55,50,50,
                        45,10,0,0,0],dtype="int32")
    })
df = pd.concat([df1 , df2] , ignore_index = True)
df
```

Some of the essential basic functionality of Pandas will now be illustrated by means of example. The reader should execute the commands in order to fully understand what the functions do.

Python Command Lines	Comments
In[1]: df.dtypes	# Lists data types of columns.
In[2]: df.head()	# Lists the first 5 rows of df.
In[3]: df.head(10)	# Lists the first 10 rows of df.
In[4]: df.tail(3)	# Lists the last 3 rows of df.
In[5]: df.index	# Lists the range index.
In[6]: df.columns	# Lists the column headings.
In[7]: df.describe	# Statistics of numerical columns.
In[8]: df["Dosage (mg)"]	# Select a single column.
In[9]: df.sort_values(by="Weight (Kg)"	# Sort data by one column.
In[10]: df.loc[1 : 3]	# Slice rows.
In[11]: df[12 :]	# Slice rows.
In[12]: df.iloc[19]	# Lists data in index 19.
In[13]: df.iloc[1 : 3 , 0 : 2]	# Slice rows and columns using index.
In[14]: df.iat[2 , 2]	# Data in row 3, column 3.
In[15]: df[df["Weight (Kg)"]>60]	# Weights bigger than 60kg.
In[16]: df.to_csv("Drug_Trial.csv")	# Writes to csv file.
In[17]: pd.csv_read("Drug_Trial.csv")	# Loads csv file.
In[18]: df.to_excel("Drug_Trial.xlsx")	# Writes to excel file.
In[19]: df.T	# Transpose the data.
In[20]: df.dropna()	# Drop rows with NaN entries.

8.2 LINEAR PROGRAMMING

Linear programming is an optimization technique in mathematics where linear functions are subject to certain constraints. The methods are used a lot in manufacturing, engineering, finance, food and agriculture, transportation, and energy, for example.

For more details, readers are directed to [8] and [11]. Some of the methods used can be illustrated with simple examples.

Example 8.2.1. An electronics manufacturer has a stock of 1500 capacitors, 2000 resistors and 1200 transistors. They produce two types of circuit. Circuit A requires 10 capacitors, 20 resistors and 10 transistors, whilst circuit B requires 30 capacitors, 10 resistors and 20 transistors. The profit on one circuit A is 5 dollars and the profit on one circuit B is 12 dollars. Plot a feasibility region and determine how many of each circuit the company should produce in order to maximize profit?

Solution. Suppose that the number of circuits of types A and B is x and y, respectively. Then the profit, or objective, function P in dollars, say, can be expressed as:

$$P = 5x + 12y. \tag{8.1}$$

The constraining equations, due to the limitations on the numbers of capacitors, resistors, and transistors, are

$$10x + 30y \leq 1500, \quad 20x + 10y \leq 2000, \quad 10x + 20y \leq 1200, \tag{8.2}$$

and

$$x \geq 0, \quad y \geq 0. \tag{8.3}$$

Program_8b.py is used to plot a feasibility region indicating the set of all possible feasible solutions, using equations (8.2) and (8.3). Figure 8.2 shows the feasibility region shaded in grey and the vertices are at $A = (0, 50)$, $B = (60, 30)$, $C = (93\frac{1}{3}, 13\frac{1}{3})$ and $D = (100, 0)$. These points may be easily computed in Python. To maximize the profit function P in equation (8.1), simply check the value of P as the line $P = 5x + 12y$ passes through the vertices A, B, C and D. The maximum profit occurs when the line passes through the point B, and is given as $P_{max} = 5(60) + 12(30) = 660$ dollars.

```
# Program_8b.py: Plotting the feasibility region and maximizing profit.
# Maximize P = 5x+12y, given 20x+10y<=2000, 10x+20y<=1200,10x+30y<=1500.
# x,y >= 0.
import numpy as np
import matplotlib.pyplot as plt
m = np.linspace(0,200,200)
x , y = np.meshgrid(m , m)
plt.imshow(((x>=0) & (y>=0) & (20*x+10*y<=2000) & (10*x+20*y<=1200) \
          & (10*x+30*y<=1500)).astype(int) , extent=(x.min(),x.max(), \
          y.min(),y.max()),origin="lower",cmap="Greys",alpha=0.3)

# Plot the constraint lines.
x = np.linspace(0, 200, 200)
y1 = (-20*x+2000) / 10
```

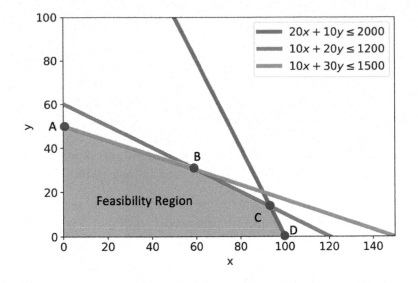

Figure 8.2 Linear programming, plotting the feasibility region for Example 8.2.1. The vertices, A-D, in this case, indicate the intersections of the bounding lines and are used to determine a maximum value, in this case.

```
y2 = (-10*x+1200) / 20
y3 = (-10*x+1500) / 30
plt.rcParams["font.size"] = "14"
plt.plot(x, y1, label=r"$20x+10y \leq 2000$" , linewidth = 4)
plt.plot(x, y2, label=r"$10x+20y \leq 1200$" ,  linewidth = 4)
plt.plot(x, y3, label=r"$10x+30y \leq 1500$" , linewidth = 4)
# The maximum profit line.
# plt.plot(x , (-5 * x + 660) / 12 , label = r"$P_{max}=5x+12y$")
plt.xlim(0,150)
plt.ylim(0,100)
plt.xlabel("x")
plt.ylabel("y")
plt.legend()
plt.show()
```

The simplex method was invented by George Dantzig in the late 1940s [2] to solve linear programs. These problems can be solved by hand using slack variables, tableuas, and pivot varaibles, for more details, see [3], for example. In Python, these problems can be solved using the linprog function from the scipy library. The first example is a minimization problem.

Example 8.2.2. A small petroleum company owns two refineries. Refineries A and B have operations costs of 20,000 dollars and 25,000 dollars per day, respectively. Refinery A produces 400, 300, and 200 barrels of high-grade, medium-grade, and low-grade oils, respectively. The corresponding figures for refinery B are 300, 400, and 500 barrels. The company has orders totalling 25,000, 27,000, and 30,000 barrels of high-grade, medium-grade, and low-grade oil, respectively. Write a Python program to work out the minimum cost, and indicate any surplus production.

Solution. Suppose that x and y are the number of days to run refineries A and B, respectively. For this problem, we need to minimize the cost (or objective) function:

$$C = 20,000x + 25,000y, \tag{8.4}$$

subject to the constraints:

$$400x + 300y \geq 25000, \quad 300x + 400y \geq 27000, \quad 200x + 500y \geq 30000, \tag{8.5}$$

and

$$x \geq 0, \quad y \geq 0. \tag{8.6}$$

Program_8c.py lists the code to solve this optimization problem. Note that the **linprog** function only solves minimization problems, where the constraints involve inequalities with less than or equal to only. Note that: $400x + 300y \geq 25000$ is equivalent to $-400x - 300y \leq -25000$, for example. Using equations (8.4) to (8.6), the output from the program is listed below. The minimum cost of 1,750,000 dollars is achieved (see fun:) when refineries A and B operate for 25 and 50 days (see x: array), respectively. There is a surplus of 500 barrels of medium-grade oil (see slack).

```
# Program_8c.py: Simplex Method to Minimize Cost.
from scipy.optimize import linprog
obj = [20000, 25000]          # Minimize C = 20000x + 25000y.
lhs_ineq = [[ -400,  -300],   # 400x + 300y.
              [-300 , -400],   # 300x + 400y
              [ -200, -500]]   #  200x + 500y.
rhs_ineq = [-25000,    # Constraint inequality 1.
              -27000,
              -30000]  # Constraint inequality 2.
bnd = [(0, float("inf")), (0, float("inf")) ]  # Bounds.
opt = linprog(c=obj, A_ub=lhs_ineq, b_ub=rhs_ineq, bounds=bnd,
        method="revised simplex")
opt
```

```
Solution output:
con: array([], dtype=float64)
     fun: 1750000.0
 message: 'Optimization terminated successfully.'
     nit: 3
   slack: array([  0., 500.,    0.])
  status: 0
 success: True
       x: array([25., 50.])
```

Example 8.2.3. Use the linprog function to solve the following optimization problem. Maximize the objective function:

$$z = x_1 - x_2 + 3x_3. \tag{8.7}$$

subject to the constraints:

$$x_1 + x_2 \leq 20, \quad x_2 + x_3 \geq 10, \quad x_1 + x_3 = 5, \tag{8.8}$$

and

$$x_1 \geq 0, \quad x_2 \geq 0, \quad x_3 \geq 0. \tag{8.9}$$

Solution. Maximizing z in equation (8.7) is equivalent to minimizing $-z = -x_1 + x_2 - 3x_3$. Program_8d.py lists the program for solving this optimization problem with constraints listed in equations (8.8) and (8.9). The output is listed below the program. The maximum value of z is $z = 10$, and this occurs when $x_1 = 0$, $x_2 = 5$ and $x_3 = 5$.

```
# Program_8d.py: Simplex method for a maximize problem.
from scipy.optimize import linprog
obj = [-1, 1, -3]            # Minimize -P = -x1 + x2 - 3x3.
                             # Maximize P = x1 - x2 + 3x3.
lhs_ineq = [[ 1,   1, 0],    # x1 + x2.
            [ 0, -1, -1]]    # - x2 - x3.
rhs_ineq = [20,              # Constraint inequality 1.
            -10]             # Constraint inequality 2.
lhs_eq = [[1, 0, 1]]         # x1 + x3.
rhs_eq = [5]                 # Constraint equation.
# Bounds of x1 , x2 , x3.
bnd = [(0, float("inf")), (0, float("inf")) , (0, float("inf"))]
opt = linprog(c=obj, A_ub=lhs_ineq, b_ub=rhs_ineq,
        A_eq=lhs_eq, b_eq=rhs_eq, bounds=bnd,
        method="revised simplex")
opt
```

```
Solution output:
con: array([0.])
      fun: -10.0
  message: 'Optimization terminated successfully.'
      nit: 2
    slack: array([15.,   0.])
   status: 0
  success: True
        x: array([0., 5., 5.])
```

8.3   K-MEANS CLUSTERING

K-means clustering is an unsupervised learning method which is part of the sci-kit-learn (**sklearn**) data analysis library in Python. The URL for scikit-learn is

```
https://scikit-learn.org.
```

Amongst the many applications are classification, regression, clustering, dimensionality reduction, model selection and preprocessing, for example. This section of the chapter will look at a simple clustering application using the k-means algorithm, where the data is split into k non-overlapping subgroups (clusters) where the distances from the data points to the centroids of the clusters is a minimum. The reader is directed to the books [1] and [4] for theory, algorithms and applications.

Example 8.3.1. Load the Boston housing data through Keras and investigate the clustering of data on the first 100 data points from column five of the data (nitrous oxide concentrations).

Solution. Program_8e.ipynb loads the Boston-housing data from column five and applies the elbow method to determine the number of possible clusters in the data with respect to the sum of the squared errors. Figure 8.3(a) shows the distribution of the data points, and Figure 8.3(b) displays the number of clusters against the sum of the squares of the errors.

```
# Program_8e.ipynb: K-means clustering and the elbow method.
# Run program in Google Colab.
import tensorflow as tf
from tensorflow import keras
import numpy as np
import matplotlib.pyplot as plt
from keras.datasets import boston_housing
(x_train,y_train)=boston_housing.load_data(path="boston_housing.npz",\
                    test_split=0.2,seed=113)
num_houses , datacol = 100 , 5
plt.scatter(x_train[0][:,datacol][1:num_houses],y_train[1][1: \
            num_houses],marker="+")
plt.rcParams["font.size"] = "18"
plt.xlabel("Nitrous Oxide Concentration (Column 5)")
plt.ylabel("Value of Houses")
plt.show()
X  = np.vstack((x_train[0][:,datacol][1:num_houses] , y_train[1] \
     [1:num_houses])).T
import seaborn as sns
from sklearn.cluster import KMeans
```

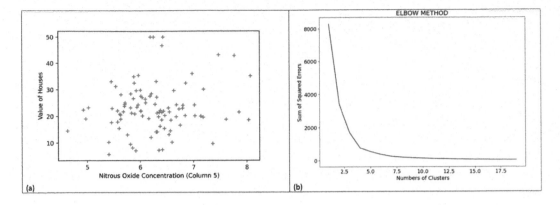

Figure 8.3 (a) Data from column five of the Boston-housing data. (b) The elbow method to determine the number of clusters.

```
elbow=[]
for i in range(1, 20):
    kmeans = KMeans(n_clusters=i,init="k-means++",random_state=101)
    kmeans.fit(X)
    elbow.append(kmeans.inertia_)
sns.lineplot(range(1, 20), elbow,color="blue")
plt.rcParams.update({"figure.figsize":(10,7.5), "figure.dpi":100})
plt.title("ELBOW METHOD")
plt.xlabel("Numbers of Clusters")
plt.ylabel("Sum of Squared Errors")
plt.show()
```

Program_8f.ipynb lists the mean shift clustering algorithm program for determining the number of clusters and Figure 8.4(a) shows the 6 cluster sets. Program_8g.ipynb plots the envelope curves for the empirical covariance, robust covariance and one-class support vector machines (OCSVM) to detect rare events. Figure 8.4(b) shows the outlier detection on a real data set of column five of the Boston housing data.

```
# Program_8f.ipynb: Mean Shift Clustering Algorithm.
import tensorflow as tf
from tensorflow import keras
import numpy as np
import matplotlib.pyplot as plt
from keras.datasets import boston_housing
from sklearn.cluster import MeanShift, estimate_bandwidth
from sklearn.datasets import make_blobs
(x_train,y_train)=boston_housing.load_data(path="boston_housing.npz", \
                    test_split=0.2,seed=113)
num_houses , datacol = 100 , 5
plt.scatter(x_train[0][:,datacol][1:num_houses],y_train[1][1: \
            num_houses],marker="+")
```

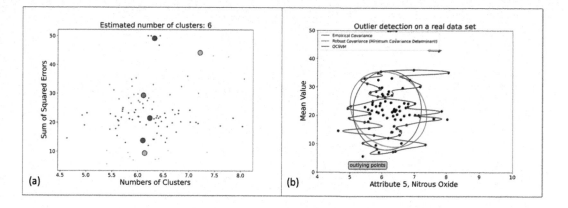

Figure 8.4 (a) Six cluster sets from the mean shift clustering algoritm. (b) Envelope curves for empirical covariance, robust covariance and OCSVM, to detect rare events.

```python
# The following bandwidth can be automatically detected using
X  = np.vstack((x_train[0][:,datacol][1:num_houses] , \
        y_train[1][1:num_houses])).T
bandwidth = estimate_bandwidth(X, quantile=0.2, n_samples=100)
ms = MeanShift(bandwidth=bandwidth, bin_seeding=True)
ms.fit(X)
labels = ms.labels_
cluster_centers = ms.cluster_centers_
labels_unique = np.unique(labels)
n_clusters_ = len(labels_unique)
print("number of estimated clusters : %d" % n_clusters_)
import matplotlib.pyplot as plt
from itertools import cycle
plt.figure(1)
plt.clf()
colors = cycle("bgrcmykbgrcmykbgrcmykbgrcmyk")
for k, col in zip(range(n_clusters_), colors):
    my_members = labels == k
    cluster_center = cluster_centers[k]
    plt.plot(X[my_members, 0], X[my_members, 1], col + ".")
    plt.plot(
        cluster_center[0],
        cluster_center[1],
        "o",
        markerfacecolor=col,
        markeredgecolor="k",
        markersize=14,)
fs = 20
plt.title("Estimated number of clusters: %d" % n_clusters_,fontsize=fs)
plt.xlabel("Numbers of Clusters",fontsize = fs)
```

```python
plt.ylabel("Sum of Squared Errors",fontsize = fs)
plt.show()
```

```python
# Program_8g.ipynb: Elliptic envelopes and OCSVM.
# Run Program_8f.ipynb in same notebook first.
import numpy as np
from sklearn.covariance import EllipticEnvelope
from sklearn.svm import OneClassSVM
import matplotlib.pyplot as plt
import matplotlib.font_manager
# Define "classifiers" to be used
classifiers = {
    "Empirical Covariance": EllipticEnvelope(support_fraction=1.0, \
    contamination=0.25),
    "Robust Covariance (Minimum Covariance Determinant)": \
    EllipticEnvelope(contamination=0.25),
    "OCSVM": OneClassSVM(nu=0.25, gamma=0.35),
}
colors = ["r", "g", "b"]
legend1 = {}
legend2 = {}
# Get data
X1 = np.vstack(((x_train[0][:,datacol][1:num_houses] , \
        y_train[1][1:num_houses])).T
# Learn a frontier for outlier detection with several classifiers
xx1,yy1=np.meshgrid(np.linspace(4,10,500),np.linspace(0,50,500))
for i, (clf_name, clf) in enumerate(classifiers.items()):
    plt.figure(1)
    clf.fit(X1)
    Z1 = clf.decision_function(np.c_[xx1.ravel(), yy1.ravel()])
    Z1 = Z1.reshape(xx1.shape)
    legend1[clf_name] = plt.contour(
        xx1, yy1, Z1, levels=[0], linewidths=2, colors=colors[i]
    )
legend1_values_list = list(legend1.values())
legend1_keys_list = list(legend1.keys())
# Plot the results (= shape of the data points cloud)
plt.figure(1)  # two clusters
plt.title("Outlier detection on a real data set (Boston Housing)")
plt.scatter(X1[:, 0], X1[:, 1], color="black")
bbox_args = dict(boxstyle="round", fc="0.8")
arrow_args = dict(arrowstyle="->")
plt.annotate(
    "outlying points",
    xy=(5, 2),
```

```
            xycoords="data",
            textcoords="data",
            xytext=(5, 2),
            bbox=bbox_args,
            arrowprops=arrow_args,
)
plt.xlim((xx1.min(), xx1.max()))
plt.ylim((yy1.min(), yy1.max()))
plt.legend(
    (
            legend1_values_list[0].collections[0],
            legend1_values_list[1].collections[0],
            legend1_values_list[2].collections[0],
    ),
    (legend1_keys_list[0],legend1_keys_list[1],legend1_keys_list[2]),
    loc="upper left",
    prop=matplotlib.font_manager.FontProperties(size=11),
)
fs = 20
plt.title("Outlier detection on a real data set", fontsize=fs)
plt.xlabel("Attribute 5, Nitrous Oxide",fontsize = fs)
plt.ylabel("Mean Value",fontsize = fs)
plt.show()
```

8.4 DECISION TREES

A decision tree has a tree like-structure (upside down) and can be used to make predictions. In the simple example below, the classification decision tree is used to predict a species of iris given its petal length and petal width. There are three species in the iris data set including setosa, versicolor, and virginica. Note that the dataset also includes data for sepal length and sepal width, but including this data would give a much larger decision tree.

Example 8.4.1. Construct a simple decision tree for the sci-kit learn iris data set and show how it can be used to predict the species of iris given some length features. Illustrate how well the decision tree works by plotting a confusion matrix.

Solution. Program_8h.ipynb is used to plot the decision tree in Figure 8.5. Note that True values always go left and False values go right. The numbers in the inequalities are computed using an optimization algorithm. Using the decision tree, it is easy to predict the species of iris given some data. For example, given [sepal length, sepal width, petal length, petal width]= [6, 4, 4, 0.4]. From the first layer (root), go right as the petal length is greater than 2.45 cm. Next, take the True path (left) as petal length is less than 4.75 cm. The petal width is less than 1.65 cm, so we go left again, and end up in the leaf of the tree with class=versicolor. The gini index ranges from

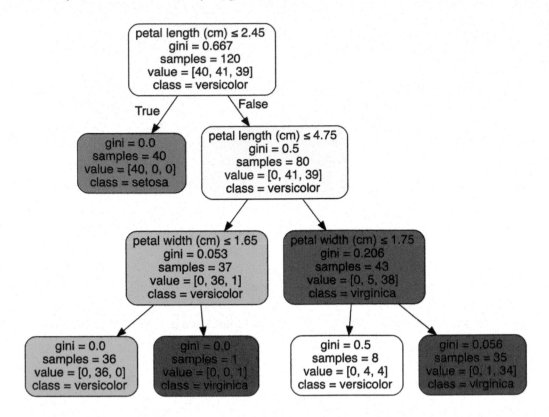

Figure 8.5 Decision tree for sklearn iris data set. One can classify an iris using only the petal length and petal width.

zero to one. A zero value indicates a purity of classification. Mathematically, the gini index, gi say, is defined by $gi = 1 - \sum_{i=1}^{n} (p_i)^2$, where p_i is the probability of an element being classified for a distinct class. For a beginners guide to decision trees, see [10], and for a visual guide to machine learning with random forests and decision trees, see [6].

```
# Program_8h.ipynb: Creating decision trees from a data frame and
# predicting output. Plotting a confusion matrix.
# Run the program in Google Colab.
import pandas as pd
import numpy as np
from sklearn import tree
from sklearn.datasets import load_iris
import graphviz
from sklearn import metrics
import seaborn as sns
import matplotlib.pyplot as plt

data = load_iris()
df = pd.DataFrame(data.data, columns = data.feature_names)
```

```python
df["Species"] = data.target
target = np.unique(data.target)
target_names = np.unique(data.target_names)
targets = dict(zip(target, target_names))
df["Species"] = df["Species"].replace(targets)
x = df.drop(columns="Species")
y = df["Species"]
feature_names = x.columns
labels = y.unique()
# Create training and testing data.
from sklearn.model_selection import train_test_split
X_train, test_x, y_train, test_lab = \
train_test_split(x,y,test_size = 0.2,random_state = 42)
clf = tree.DecisionTreeClassifier(max_depth = 3, random_state = 42)
clf.fit(X_train, y_train)
dot_data = tree.export_graphviz(clf, out_file=None, \
                                feature_names=data.feature_names, \
                                class_names=data.target_names, \
                                filled=True, rounded=True, \
                                special_characters=True)
DTclf=clf.fit(X_train, y_train)
prediction = DTclf.predict([[6,4,4,0.4]])
print("Prediction for input [6,4,4,0.4]:" , prediction)
graph = graphviz.Source(dot_data)
# Plot a Confusion Matrix.
plt.figure(figsize=(10,6))
test_pred_DT = clf.predict(test_x)
confusion_matrix = metrics.confusion_matrix(test_lab, test_pred_DT)
matrix_df = pd.DataFrame(confusion_matrix)
ax = plt.axes()
sns.set(font_scale=2)
sns.heatmap(matrix_df, annot=True, fmt="g", ax=ax, cmap="viridis")
ax.set_title("Confusion Matrix - Decision Tree")
ax.set_xlabel("Predicted label", fontsize =20)
ax.set_xticklabels([""]+labels)
ax.set_ylabel("True Label", fontsize=20)
ax.set_yticklabels(list(labels), rotation = 0)
plt.show()
# Plot the decision tree.
graph
```

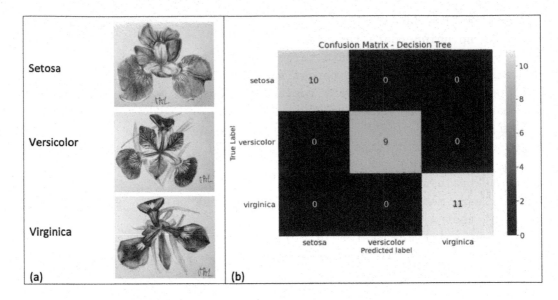

Figure 8.6 (a) Pictures of the three species of iris, hand-drawn by my daughter Thalia Lynch. (b) Confusion matrix for the decision tree plotted in Figure 8.5. A confusion matrix is a table used to define the performance of a classification algorithm. In this case, the performance looks perfect.

EXERCISES

8.1 Use Python to create a data frame for the attendance numbers and satisfaction ratings at a gymnasium in Figure 8.7. Make the "Date" column the index.

8.2 (a) Using the same parameters in Example 8.2.2, plot the feasibility region for this optimization problem.

(b) Use Python and the function **linprog** to maximize the function:

$$z = 20x_1 + 10x_2 + 15x_3,$$

subject to the constraints:

$$3x_1 + 2x_2 + 5x_3 \leq 55,$$

$$2x_1 + x_2 + x_3 \leq 26,$$

$$x_1 + x_2 + 3x_3 \leq 30,$$

$$5x_1 + 2x_2 + 4x_3 \leq 57,$$

where

$$x_1, x_2, x_3 \geq 0.$$

Date	Day	Member Numbers	Staff Numbers	Member/Staff Ratio	Satisfaction (%)
2022-01-09	Sunday	126	12	10.5	34.0
2022-01-10	Monday	34	6	5.7	58.0
2022-01-11	Tuesday	42	6	7.0	74.0
2022-01-12	Wednesday	100	12	8.3	54.0
2022-01-13	Thursday	54	6	9.0	85.0
2022-01-14	Friday	41	6	6.8	88.0
2022-01-15	Saturday	105	12	8.8	45.0

Figure 8.7 Pandas data frame for satisfaction ratings at a gymnasium over a one-week period. The "Date" column acts as the index.

8.3 Carry out the same k-means clustering methods as in Example 8.3.1 for the column 12 (pupil to teacher ratio by town) data of the Boston housing data. What can you conclude?

8.4 Load the csv file, "possum.csv" from GitHub. Create a regression decision tree using only the columns entitled "hdlngth" for head length, "skullw" for skull width, and "totlngth" for total length, to predict the age of the possum. Take a maximum depth of three for the decision tree and split the training and testing data in the ratio 70:30.

Solutions to the Exercises may be found here:

https://drstephenlynch.github.io/webpages/Solutions_Section_2.html.

FURTHER READING

[1] Aggarwal, C.C. and Reddy, C.K. (Editors). (2013). *Data Clustering: Algorithms and Applications.* CRC Press, Boca Raton, FL.

[2] Dantzig, G.B. (1948). Linear Programming, *in Problems for the Numerical Analysis of the Future, Proceedings of the Symposium on Modern Calculating Machinery and Numerical Methods, UCLA (July 29-31); Appl. Math.* 15, National Bureau of Standards (1951): 18-21.

[3] Ficken, F.A. (2015). *The Simplex Method of Linear Programming.* Dover Publications, New York.

[4] Gan, G., Ma, C. and Wu, J. (2020). *Data Clustering: Theory, Algorithms, and Applications, 2nd Ed.* SIAM, Philadelphia.

[5] Grus, J. (2019). *Data Science from Scratch: First Principles with Python, 2nd Ed.* O'Reilly Media, Sebastopol, CA.

[6] Hartshorn, S. (2016). *Machine Learning With Random Forests And Decision Trees: A Visual Guide For Beginners.* Kindle Store.

[7] Klosterman, S. (2021). *Data Science Projects with Python: A case study approach to gaining valuable insights from real data with machine learning, 2nd Ed.* Packt Publishing, Birmingham.

[8] Matousek, J. and Gartner, B. (2006). *Understanding and using Linear Programming.* Springer, New York.

[9] McKinney, W. (2017). *Python for Data Analysis: Data Wrangling with Pandas, NumPy, and IPython, 2nd Ed..* O'Reilly Media, Sebastopol, CA.

[10] Smith, C. and Koning, M. (2017). *Decision Trees and Random Forests: A Visual Introduction For Beginners: A Simple Guide to Machine Learning with Decision Trees.* Blue Windmill Media.

[11] Vanderbel, R.J. (2020). *Linear Programming: Foundations and Extensions.* Springer, New York.

[12] VanderPlas, J. (2017). *Python Data Science Handbook: Essential Tools for Working with Data.* O'Reilly Media, Sebastopol, CA.

## CHAPTER 9

Economics

Economics is the science of production, consumption and transfer of wealth. The economics of single factors and the effects of individual decisions is classified as microeconomics, whereas large-scale factors, such as inflation, interest rates, national productivity and unemployment, for example, are classed as macroeconomics, and deal with the economy as a whole. The first example is from microeconomics and presents a simple example of quantity of production. We use the Lagrange multiplier technique to maximize output subject to restrictions on cost. The second example is taken from macroeconomics and is known as the Solow-Swan exogenous growth model that monitors changes in economic output over time. The model consists of a single ODE that models the evolution of the per capita stock of capital. An interactive plot is created to see how solution curves vary as parameters in the model change. The third model is taken from modern portfolio theory (MPT), the Markowitz model is used to extract efficient portfolios by analyzing random portfolio combinations based on expected returns (means) and standard deviations (variances) of assets. The final example presents the Black-Scholes model used to determine the dynamics of a financial market, in particular, the price of options over time. The Greeks are quantities representing the sensitivities of the price of derivatives, such as options, with respect to underlying stock price, time to expiration, and the implied volatility of the underlying stock. The model is created from scratch and the results can be compared to online models.

Recently, there have been new and exciting developments in the field of economics and financial mathematics, in 2020, Bukhari et al. [3] reported an 80% improvement on forecasting using Artificial Intellegence (AI) with deep learning using long-short-term-memory, and in 2018, Drozdz et al. [6] used variety in multifractal curves to show varying degrees of correlations among stocks.

Aims and Objectives:

- To provide examples of mathematical models in economics.

At the end of the chapter, the reader should be able to

- solve optimization problems using Lagrange multipliers,

- use an interactive plot to see how parameters affect a model;

DOI: 10.1201/9781003285816-9

- use the Black-Scholes model to determine the fair price, or theoretical value, for a call or put option based on the underlying stock price, time to expiration, and implied volatility of the underlying stock.

9.1 THE COBB-DOUGLAS QUANTITY OF PRODUCTION MODEL

In 1928, Cobb and Douglas [5] modeled the relationship between quantity of product and cost for a manufacturer. Consider the following simple example:

$$Q(L, K) = AL^{\alpha}K^{1-\alpha}, \quad C(L, K) = wL + rK,$$

where Q is the quantity of a product based on L and K, representing labor and capital, respectively, α and $(1 - \alpha)$, are the output elasticities of labor and capital, respectively, and C is the cost of construction, where w is the per hour labor costs, and r per tonne product costs. Cobb and Douglas described how to maximize quantity of product using a given cost, equivalently, you can minimize $C(L, K)$ given a function $Q(L, K)$. This is an optimization problem that can be solved using the method of Lagrange multipliers [8], which is defined here.

Definition 9.1.1. To maximize (or minimize) a multivariable function $f(x_1, x_2, \ldots, x_n)$, subject to a constraint $g(x_1, x_2, \ldots, x_n) = 0$, set a Lagrangian function:

$$L(x_1, x_2, \ldots, x_n, \lambda) = f(x_1, x_2, \ldots, x_n) - \lambda g(x_1, x_2, \ldots, x_n),$$

and solve the following set of $n + 1$ simultaneous equations in $n + 1$ unknowns:

$$\frac{\partial L}{\partial x_1} = \frac{\partial L}{\partial x_2} = \cdots = \frac{\partial L}{\partial x_n} = 0, \quad g(x_1, x_2, \ldots, x_n) = 0.$$

The solutions to these equations determine the maximum (or minimum) values that you seek.

Example 9.1.1. Consider a US manufacturer of steel cans, the functions $Q(L, K)$ and $C(L, K)$, are given by:

$$Q(L, K) = 200L^{\frac{2}{3}}K^{\frac{1}{3}}, \quad C(L, K) = 20L + 750K, \quad (9.1)$$

where the cost of labor, w, is 20 dollars per hour, and the steel costs, r, are 750 dollars per tonne. Plot a graph showing isocost lines (showing all combinations of inputs which cost the same total amount) and isoquant curves (a contour drawn through the set of points at which the same quantity of output is produced while changing the quantities of L and K), for the cost functions (i) $C_1 : 20L + 750K = 10000$, and (ii) $C_2 : 20L + 750K = 20000$.

Solution. Program_9a.py produces the isocost-isoquant graph shown in Figure 9.1. It uses the Lagrange multiplier method to determine the tangential points. The curve through the tangential points is known as the expansion path.

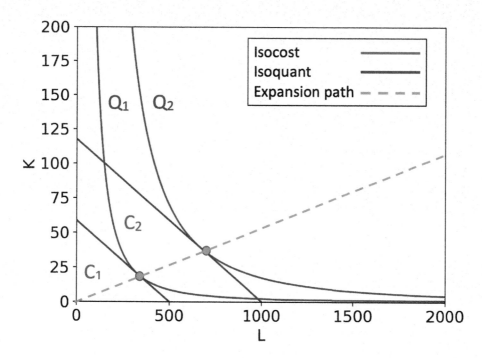

Figure 9.1 An isocost-isoquant graph for a Cobb-Douglas model (8.1) of quantity of product. The given cost constraints are, $C_1 : 20L + 750K = 10000$, and $C_2 : 20L + 750K = 20000$. The expansion path is a curve connecting optimal combinations as the scale of production expands. For the Cobb-Douglas model, the curve is a straight line through the origin.

```
# Program_9a.py: Cobb-Douglas Model of Production.
import numpy as np
import matplotlib.pyplot as plt
from sympy import symbols , diff , solve
L,K,lam=symbols("L K lam")
Lmax , Kmax  = 2000 , 200
w , r = 20 , 170
Y = 200*L**(2/3)*K**(1/3)
C=10000
Lagrange=Y-lam*(w*L+r*K-C)
L1 = diff(Lagrange,L)
L2 = diff(Lagrange,K)
L3 = w*L+r*K-C
sol=solve([L1,L2,L3],L,K,lam)
Y1 = 200*sol[0][0]**(2/3)*sol[0][1]**(1/3)
C=20000
Lagrange=Y-lam*(w*L+r*K-C)
L1 = diff(Lagrange,L)
L2 = diff(Lagrange,K)
```

```
L3 = w*L+r*K-C
sol=solve([L1,L2,L3],L,K,lam)
Y2 = 200*sol[0][0]**(2/3)*sol[0][1]**(1/3)
Llist = np.linspace(0,Lmax, 1000)
Klist = np.linspace(0, Kmax, 120)
L, K = np.meshgrid(Llist, Klist)
plt.figure()
Z = 200*L**(2/3)*K**(1/3)
plt.contour(L,K,Z,[Y1,Y2],colors="red")
Z = 20*L+170*K
plt.contour(L,K,Z,[10000,20000],colors="blue")
plt.xlabel("L",fontsize=15)
plt.ylabel("K",fontsize=15)
plt.tick_params(labelsize=15)
plt.show()
```

9.2 THE SOLOW-SWAN MODEL OF ECONOMIC GROWTH

The neoclassical Solow-Swan growth model is used to study the exogenous impact of capital share, depreciation rate, population growth rate, savings rate, technological change and labor participation rate, on economic growth. The ODE is

$$\frac{dk(t)}{dt} = s(f(k(t))) - (n + \delta + g)k(t), \tag{9.2}$$

where $k(t)$ is capital intensity (the capital stock per unit of effective labor), the term $s(f(k(t))$, is the actual investment per unit of effective labor, s is the savings rate, the second term, $(n + \delta + g)k(t)$, is the break-even investment, representing the amount of investment required to prevent k from falling, and n, δ, g, represent population growth, depreciation, and technological progress, respectively. We employ the standard Cobb-Douglas production function, $f(k(t)) = k(t)^\alpha$, where $0 < \alpha < 1$, is the capital share, in this model. In a recent paper, Cayssials and Picasso [4] improve the 1950's Solow and Swan models [10], [11], by incorporating the dynamics of an endogenous population.

Example 9.2.1. Create an interactive plot (see Program_3b.py) for the solution curves of ODE (9.1) for parameter values, $0 < \alpha < 1$, $0 < \delta < 1$, $0 < g < 0.1$, $0 < n < 0.1$ and $0 < s < 0.5$.

Solution. Program_9b.py produces the interactive plot shown in Figure 9.2.

```
# Program_9b.ipynb: The Solow-Swan Model of Economic Growth.
# Run program in Google Colab.
from __future__ import print_function
from ipywidgets import interact, interactive, fixed, interact_manual
```

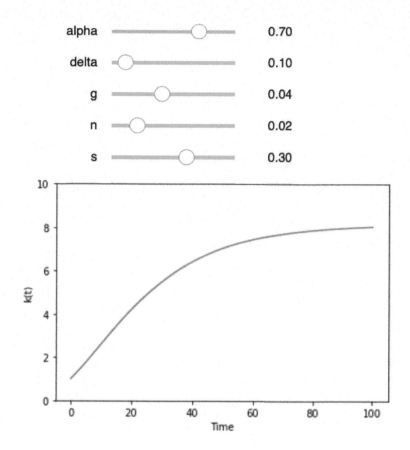

Figure 9.2 An interactive plot for the ODE (9.1) when $\alpha = 0.7$, $\delta = 0.1$, $g = 0.04$, $n = 0.02$ and $s = 0.3$. As the parameters vary, the capital intensity $k(t)$ reaches different steady-states.

```
import ipywidgets as widgets
%matplotlib inline
from ipywidgets import interactive
import matplotlib.pyplot as plt
import numpy as np
from scipy.integrate import odeint

Tmax , Ylim = 100 , 10
def f(alpha , delta , g , n , s):
    def Solow_Swan(k,t):
      dkdt = s * k**alpha - (n + delta + g) * k
      return dkdt
    k0 = 1
    t = np.linspace(0,Tmax)
    y = odeint(Solow_Swan,k0,t)
    plt.ylim((0,Ylim))
```

```
plt.plot(t,y)
plt.xlabel("Time")
plt.ylabel("k(t)")
plt.show()
interactive_plot=interactive(f,alpha=(0,1,0.1),delta = (0,1,0.1), \
                             g = (0,0.1,0.01), n = (0,0.1,0.01),\
                             s=(0,0.5,0.1))
output = interactive_plot.children[-1]
output.layout.height = "350px"
interactive_plot
```

9.3 MODERN PORTFOLIO THEORY (MPT)

In 1952, Markowitz [9] introduced MPT, or mean-variance analysis, as a mathematical tool for assembling a portfolio of assets such that the expected return is maximized for a given level of risk. Readers are encouraged to read the excellent book, Modern Portfolio Theory and Investment Analysis, by Elton et al. [7] for a more detailed introduction. We start by downloading financial datasets using Yahoo Finance **yfinance** in Python. In Google Colab, we **!pipinstall yfinance**. In Program_9c.py, financial data is loaded for three stocks, Apple, Caterpillar and Google using the ticker (stock) symbols, AAPL, CAT and GOOG, respectively. Next, the data is normalized, and then exported as a csv file to the Google Colab folder. Readers can then use the data in Spyder without access to the internet if they so wish.

Anyone can invest in stocks (a security that represents the ownership of a part of a company), bonds (a loan made buy an investor to a borrower), exchange traded funds (EFTs) (a pooled investment security that operates like a mutual fund), for example. Another, shorter form of investment, lies with call and put options, which are a form of derivative (financial contract) that gives the option holders the right (but not the obligation) to buy or sell a security at a chosen price at some time in the future.

The expected return for a portfolio, R_P say, with n assets is defined as:

$$E\left(R_P\right) = \sum_{i=1}^{n} w_i E\left(R_i\right),$$

where w_i and R_i are the weights and returns on assets i, respectively. The sum of all the individual components in the weight vector, $\mathbf{w}$ say, sum to one, since they are a percentage of the total capital invested:

$$\sum_{i=1}^{n} w_i = 1.$$

The variance of the portfolio is then:

$$\sigma_P^2 = \mathbf{w}^T M \mathbf{w} = (w_1 \, w_2 \, \ldots \, w_n) \begin{pmatrix} \sigma_1^2 & \sigma_{12} & \cdots & \sigma_{1n} \\ \sigma_{21} & \sigma_2^2 & \cdots & \sigma_{2n} \\ \vdots & \vdots & \vdots & \vdots \\ \sigma_{n1} & \sigma_{n2} & \cdots & \sigma_n^2 \end{pmatrix} \begin{pmatrix} w_1 \\ w_2 \\ \vdots \\ w_n \end{pmatrix},$$

where M is the covariance matrix, which is symmetric, and the diagonal elements represent the variance of the n stocks. The standard deviation of the portfolio is

$$\sigma_P = \sqrt{\sigma_P^2}.$$

Definition 9.3.1. The Sharpe-Ratio, SR say, is the average return earned in excess of the risk-free rate per unit of volatility or total risk. It is defined by

$$SR = \frac{R_P - R_f}{\sigma_P},$$

where R_P and R_f are the return of Portfolio and risk-free rate, respectively.

Using random numbers, one can construct various portfolios by changing the capital weights of the stocks in the portfolio.

Example 9.3.1. Using the data downloaded above, plot a graph of 500 random portfolios for σ_P (risk) versus R_P (return). On the graph, indicate the portfolio with the least risk and the highest Sharpe Ratio.

Solution. Program_9c.py produces the plot shown in Figure 9.3. If you are a pension fund manager trying to minimize risk, you would choose the portfolio highlighted by the red circle—the minimum risk portfolio. A hedge fund manager would choose the red star portfolio, attempting to achieve the best risk-reward combination. The red-dashed curve is known as the efficient frontier, the curve indicates the combination of investments that will provide the highest level of return for the lowest level of risk and is known to be part of a hyperbola.

```
# Program_9c.py: Modern Portfolio Theory.
!pip install yfinance
import yfinance as yf
import numpy as np
import matplotlib.pyplot as plt
df=yf.download(["AAPL","CAT","GOOG"],start="2000-12-01",\
            end="2020-12-01")
df=np.log(1+df["Adj Close"].pct_change())
df.to_csv("Portfolio_Data.csv")
def PR(weights): # Portfolio Return
  return np.dot(df.mean(),weights)
```

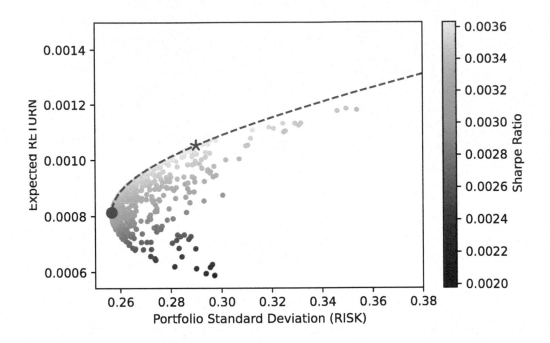

Figure 9.3 The efficient frontier (part of a hyperbola) on a RISK-RETURN graph. In this case, the Sharpe-Ratio is simply $SR = \frac{R_P}{\sigma_P}$.

```
def PSTD(weights): # Portfolio STD
  return (np.dot(np.dot(df.cov(),weights),weights))**(1/2)*\
  np.sqrt(250)
  # Annualize STD, 250 trading days.
def RW(df): # Random Weights
  rand=np.random.random(len(df.columns))
  rand /= rand.sum()
  return rand
returns , stds , w , SharpeArr = [] ,[] ,[], []
for i in range(500):
  weights=RW(df)
  returns.append(PR(weights))
  stds.append(PSTD(weights))
  w.append(weights)
  SharpeArr.append(returns[i]/stds[i])
print("Max Sharpe: {}".format(max(SharpeArr)))
print("Loc SA: {}".format(np.argmax(SharpeArr)))
argmaxSA=np.argmax(SharpeArr)
maxsa=returns[np.argmax(SharpeArr)]
maxstds=stds[np.argmax(SharpeArr)]
```

```
plt.scatter(stds,returns,c=SharpeArr,cmap="viridis",s=10)
plt.colorbar(label="Sharpe Ratio")
plt.scatter(stds[argmaxSA],returns[argmaxSA],c="r",s=70,marker=(5, 2))
plt.scatter(min(stds),returns[stds.index(min(stds))],c="r",s=70)
plt.xlabel("Portfolio Standard Deviation (RISK)")
plt.ylabel("Expected RETURN")

a , h =min(stds) , returns[stds.index(min(stds))]
xx , yy = stds[argmaxSA] , returns[argmaxSA]
b=np.sqrt((a**2*(yy-h)**2)/(xx**2-a**2))

xlist = np.linspace(0.25,0.38, 1000)
y1list = np.linspace(h, 0.0015, 120)
x, y1 = np.meshgrid(xlist, y1list)
y2list = np.linspace(0.0004, h, 120)
x, y2 = np.meshgrid(xlist, y2list)
Z = x**2/a**2-(y1-h)**2/b**2
plt.contour(x,y1,Z,[1],colors="red",linestyles="dashed")
#Z = x**2/a**2-(y2-h)**2/b**2
#plt.contour(x,y2,Z,[1],colors="blue")
plt.savefig("Fig9.3.eps" , dpi = 400)
plt.show()
```

9.4 THE BLACK-SCHOLES MODEL

The Black-Scholes equation, developed in 1973 [2] has long been regarded as one of the best ways for pricing an options contract, however, with the advent of Artificial Intelligence, new and improved models are now coming to the fore [1]. The second-order partial differential equation is written as:

$$\frac{\partial V}{\partial t} + \frac{1}{2}\sigma^2 S^2 \frac{\partial^2 V}{\partial S^2} + rS\frac{\partial V}{\partial S} - rV = 0,$$

where $V(S,t)$ is the price of the option, σ is the standard deviation of the stock's returns and is a measure of volatility, $S(t)$ is the price of the underlying asset at time t in years, and r is the annualized risk-free interest rate continuously compounded."

Black and Scholes found the solution by applying the boundary conditions $C(0,t) = 0$, for all t, $C(S,t) \to S$, as $S \to \infty$, and $C(S,t) = \max(S - K, 0)$, where $C(S,t)$, is the price of a European call option, and K is the strike price of the option. They wrote:

$$
\begin{aligned}
\text{Call} &= S_0 N(d_1) - N(d_2) Ke^{-rT}, \\
\text{Put} &= N(-d_2) Ke^{-rT} - N(-d_1) S_0, \\
d_1 &= \frac{1}{\sigma\sqrt{T}}\left(\ln\left(\frac{S}{K}\right) + \left(r + \frac{\sigma^2}{2}\right)T\right), \\
d_2 &= d_1 - \sigma\sqrt{T},
\end{aligned}
$$

(9.3)

where S_0 is the value of the call option at time zero, T is time until option expiration and $N(x)$ is the cummulative distribution function for a standard normal distribution.

Associated with the Black-Scholes solutions are the Greeks for European options, which give a measure of the sensitivities of the option prices to some parameters listed above. For an in-depth guide see [12]. They are defined by

$$\delta = \frac{\partial V}{\partial S}, \; \gamma = \frac{\partial^2 V}{\partial S^2}, \; \theta = \frac{\partial V}{\partial t}, \; \nu = \frac{\partial V}{\partial \sigma}.$$

For a call option, the delta, δ is

$$\delta_C = N(d_1), \quad 0 \leq \delta_C \leq 1,$$

and the δ for a put is

$$\delta_P = N(d_1) - 1, \quad -1 \leq \delta_C \leq 0.$$

The δ measures how much an options price will change if the underlying stock changes by one point.

For call and put, we have the gamma, γ:

$$\gamma = \frac{N'(d_1)}{S\sigma\sqrt{T}},$$

where $N'(x)$ is the probability distribution function. The γ measures how much S will change with a stock move of one point.

For call and put, we have vega, (ν):

$$\nu = 0.01SN'(d_1)\sqrt{T}.$$

The ν measures how much an options price will change if implied volatility increases by one percent.

For a call option, the theta, θ is

$$\theta_C = \frac{-SN'(d_1)\sigma}{365\left(2\sqrt{T} - rKe^{-rT}N(d_2)\right)},$$

and the θ for a put is

$$\theta_P = \frac{-SN'(d_1)\sigma}{365\left(2\sqrt{T} - rKe^{-rT}N(-d_2)\right)},$$

where θ measures how much an options price will change as one day passes.

Example 9.4.1. Use Program_9d.py to compute the option contracts for call and put, and the Greeks when $r = 0.01$, $S = 30$, $K = 40$, $T = 240/365$ and $\sigma = 0.3$.

Solution. Program_9d.py gives the output:

```
r= 0.01 S= 30 K= 40 T= 0.6575342465753424 sigma= 0.3
Option CALL price is:  0.51
Option PUT price is:  10.25
delta Call is:  0.1506
delta Put is:  -0.8494
gamma Call/Put is:  0.032
vega Call/Put is:  0.0569
theta Call is:  -0.0037
theta Put is:  -0.0045
```

```python
# Program_9d.py: Black-Scholes Option Prices for Call/Put.
# Computing the Black-Scholes Greeks.
import numpy as np
from scipy.stats import norm
# Parameters: r=interest rate,S=underlying price ($),Strike price ($),
#T=240/365 days, sigma=volatility, C=CALL, P=PUT.
r , S , K , T , sigma  = 0.01 , 30 , 40, 240/365 , 0.3
def Black_Scholes(r,S,K,T,sigma,type="C"):
  d1 = (np.log(S/K)+(r+sigma**2/2)*T)/(sigma*np.sqrt(T))
  d2 = d1 - sigma*np.sqrt(T)
  try:
    if type=="C":
      price=S*norm.cdf(d1,0,1)-K*np.exp(-r*T)*norm.cdf(d2,0,1)
      delta_calc=norm.cdf(d1,0,1)
      gamma_calc=norm.pdf(d1,0,1)/(S*sigma*np.sqrt(T))
      vega_calc=S*norm.pdf(d1,0,1)*np.sqrt(T)*0.01
      theta_calc=(-S*norm.pdf(d1,0,1)*sigma/(2*np.sqrt(T))-\
                  r*K*np.exp(-r*T)*norm.cdf(d2,0,1)) / 365
      rho_calc=K*T*np.exp(-r*T)*norm.cdf(d2,0,1)*0.01
    elif type=="P":
      price=K*np.exp(-r*T)*norm.cdf(-d2,0,1)-S*norm.cdf(-d1,0,1)
      delta_calc=-norm.cdf(-d1,0,1)
      gamma_calc=norm.pdf(d1,0,1)/(S*sigma*np.sqrt(T))
      vega_calc=S*norm.pdf(d1,0,1)*np.sqrt(T) * 0.01
      theta_calc=(-S*norm.pdf(d1,0,1)*sigma/(2*np.sqrt(T))-\
                  r*K*np.exp(-r*T)*norm.cdf(-d2,0,1)) / 365
      rho_calc=-K*T*np.exp(-r*T)*norm.cdf(-d2,0,1) * 0.01
    return [price,delta_calc,gamma_calc,vega_calc,theta_calc,rho_calc]
  except:
    print("Please input correct parameters")
BS_Call=Black_Scholes(r,S,K,T,sigma,type="C")
BS_Put=Black_Scholes(r,S,K,T,sigma,type="P")
print("r=",r,"S=",S,"K=",K,"T=",T,"sigma=",sigma)
print("Option CALL price is: ", round(BS_Call[0],2))
```

```
print("Option PUT price is: ", round(BS_Put[0],2))
print("delta Call is: ", round(BS_Call[1],4))
print("delta Put is: ", round(BS_Put[1],4))
print("gamma Call/Put is: ", round(BS_Call[2],4))
print("vega Call/Put is: ", round(BS_Call[3],4))
print("theta Call is: ", round(BS_Call[4],4))
print("theta Put is: ", round(BS_Put[4],4))
```

EXERCISES

9.1 Edit Program_9a.py to add the isocost C_3, and isoquant Q_3, curves for C_3 : $20L + 750K = 30000$, on to Figure 9.1. Determine the gradient of the expansion path line.

9.2 Consider the Solow-Swan model with the same parameters as in equation (9.2):

$$\frac{dk(t)}{dt} = s(t)k(t)^\alpha - (n + \delta + g)k(t),$$

where $s(t)$ now varies with time. Use Python to plot interactive plots when (i) $s(t) = 0.3 + 0.2\left(1 - e^{-0.1t}\right)$, and (ii) $s(t) = 0.3 + 0.2\sin(t)$. Describe what happens physically.

9.3 Edit Program_9c.py to plot the efficient frontier curve displayed in Figure 9.3.

9.4 Look on the internet for information on the vectorized py vollib package, an easy and intuitive interface for pricing thousands of option contracts and computing the Greeks. Using Google Colab, pip install this package and check the results provided by Program_9d.py.

Solutions to the Exercises may be found here:

https://drstephenlynch.github.io/webpages/Solutions_Section_2.html.

FURTHER READING

[1] Barrau, T. and Douady, R. (2022). *Artificial Intelligence for Financial Markets: The Polymodel Approach*. Springer, New York.

[2] Black, F. and Scholes, M (1973). The pricing of options and corporate liabilities. *Journal of Political Economy* **81**(3), 637–654.

[3] Bukhari, A.H., Raja, M.A.Z. Sulaiman, M. et al. (2020). Fractional neuro-sequential ARFIMA-LSTM for financial market forecasting. *IEEE Access* **8**(3), 71326–71338.

[4] Cayssials, G. and Picasso, S. (2020). The Solow-Swan model with endogenous population growth. *Journal of Dynamics and Games* **7**(3), 197–208.

[5] Cobb, C.W. and Douglas, P.H. (1928). The theory of production. *American Economic Review* **18**(3), 139–165.

[6] Drozdz, S. Kowalski, R. Oswiecima P. et al. (2018). Dynamical variety of shapes in financial multifractality. *Complexity* 70157271, 637–654.

[7] Elton, E.J., Gruber, M.J., Brown, S.J. and Goetzmann, W.N. (2014). *Modern Portfolio Theory and Investment, 9th Ed.* Wiley, New York.

[8] Lasdon, L.S. (2000). *Optimization Theory for Large Systems.* Dover Publications, New York.

[9] Markowitz, H.M. (1952). Portfolio selection. *The J. of Finance* **7**(1), 77–91.

[10] Solow, R.M. (1956). Contribution to the theory of economic growth. *The Quarterly Journal of Economics* **70**(1), 65–94.

[11] Swan, T.W. (1956). Economic growth and capital accumulation. *Economic Record* **32**(2), 334–361.

[12] Ursonwe, P (2015). *How to Calculate Options Prices and Their Greeks: Exploring the Black Scholes Model from Delta to Vega.* Wiley, New York.

## CHAPTER 10

Engineering

Engineering is the application of mathematics and science to build structures and machines. It includes subcategories such as architectural, chemical, civil, computer, electrical and mechanical, for example. This chapter will concentrate on electrical and mechanical engineering only. Electrical engineering is concerned with devices and systems that use electricity, electrons and electromagnetism—interested readers are directed to [1] and [5]. Mechanical engineering is the study of objects and systems in motion—for more detailed information, the reader is directed to [3] and [6]. The first section is on linear electrical circuits composed of resistors, inductors, capacitors and voltage sources, and introduces a fourth fundamental component, called the memristor, which was discovered in 1970 and first built in 2008. The second section introduces the reader to Chua's nonlinear electrical circuit, and a program for producing an animation is listed. The third and fourth sections cover mechanical engineering with simple examples of coupled oscillators, and periodically forced mechanical systems, respectively.

Aims and Objectives:

- To provide examples from electrical engineering.

- To investigate examples from mechanical engineering.

At the end of the chapter, the reader should be able to

- solve ODEs for simple linear electrical circuits and plot solution curves;

- understand output from nonlinear electrical circuits;

- model coupled oscillators and periodically forced mechanical systems.

10.1 LINEAR ELECTRICAL CIRCUITS AND THE MEMRISTOR

Linear electrical circuits contain components whose values do not change with the level of voltage or current in the circuit. They are important in real-world applications as they amplify and process electronic signals without distortion. Resistor-inductor-capacitor (RLC) circuits can be combined in a number of different configurations and

DOI: 10.1201/9781003285816-10

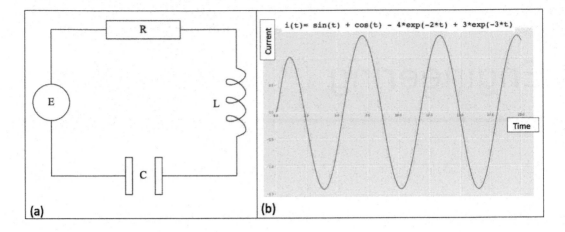

Figure 10.1 (a) A simple series resistor-inductor-capacitor (RLC) circuit. (b) The current in the circuit for equation (10.1) with $i(0) = \frac{di}{dt}(0) = 0$, $R = 5$ (Ω), $L = 1$ H, $C = \frac{1}{6}$ F and the applied voltage is $\frac{dE}{dt} = 10\cos(t)$ volts.

have applications in tuning, filtering (low-pass, high-pass, band-pass, band-stop), and can act as oscillators, voltage multipliers and pulse discharge circuits, for example. The differential equation used to model a simple series RLC circuit as depicted in Figure 10.1(a) is given by

$$L\frac{d^2i}{dt^2} + R\frac{di}{dt} + \frac{i}{C} = \frac{dE}{dt}, \tag{10.1}$$

where $i(t)$ is current, RLC give resistance, inductance and capacitance, respectively, and E is the electromotive force (voltage source).

Example 10.1.1. Given that $R = 5$ ohms (Ω), $L = 1$ henry (H), $C = \frac{1}{6}$ farads (F), the applied electromotive force (EMF) is $\frac{dE}{dt} = 10\cos(t)$ volts, and $i(0) = \frac{di}{dt}(0) = 0$, in equation (10.1), use Python to solve the ODE and plot a solution curve for the current in the circuit.

Solution. Program_10a.py, is used to solve the ODE (10.1), and plots a solution curve as shown in Figure 10.1(b). Note the use of **ggplot** to give a grid plot so it looks like the output on an oscilloscope.

```
# Program_10a.py: Current in a Resistor-Inductor-Capacitor circuit.
from sympy import symbols , diff , Eq , Function , dsolve , cos , plot
from matplotlib import style
t=symbols("t")
i=symbols("i",cls=Function)
deqn1=Eq(i(t).diff(t,t) + 5*i(t).diff(t) + 6*i(t), 10*cos(t))
odesol1=dsolve(deqn1, i(t),ics={i(0): 0, diff(i(t), t).subs(t,0): 0})
print("i(t)=",odesol1.rhs)
```

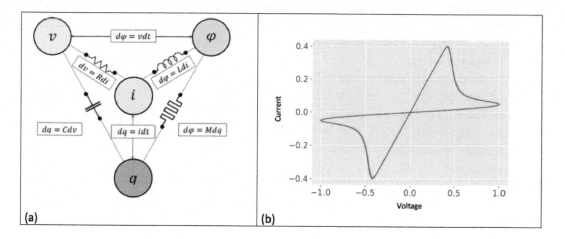

Figure 10.2 (a) Relations between voltage v, flux ϕ, current i and charge q. (b) Pinched hysteresis curve when $\omega_0 = 0.6$.

```
style.use("ggplot")
plot(odesol1.rhs , (t , 0 , 20),xlabel = "Time", ylabel = "Current")
```

For many decades, electrical engineers believed that there were only three fundamental components (RLC) in electric circuit theory. In 1970, Leon Chua, known as the grandfather of nonlinear electric circuit theory, proved the existence of a fourth component [4], which he labelled the memristor, which relates charge and flux, see Figure 10.2(a). The symbol for a memristor is indicated in Figure 10.2(a), and it acts like a nonlinear resistor which has memory. In 2008, HP labs built a titanium dioxide memristor [11] modeled using the ODE:

$$\frac{dw}{dt} = \frac{\eta f(w(t), p)v_0 \sin\left(\frac{2\pi t}{T}\right)}{w\text{R}_{\text{ON}} + (1-w)\text{R}_{\text{OFF}}}, \tag{10.2}$$

where η is the polarity of the memristor (if $\eta = +1$, then w increases with positive voltage), v_0 is the voltage amplitude, and the function, $f(w(t), p) = 1 - (2w(t) - 1)^{2p}$, is the window function for the nonlinear dopant drift. The differential equation has initial condition $w_0 = w(0)$. The differential equation (10.2) can be solved with Python and the voltage against current pinched hysteresis loop can be plotted, see Figure 10.2(b).

Example 10.1.2. Plot the pinched hysteresis curve for the HP labs titanium dioxide memristor using equation (10.2) when $\eta = 1$, $L = 1$, $\text{R}_{\text{OFF}} = 70$, $\text{R}_{\text{ON}} = 1$, $p = 10$, $T = 20$ and $\omega_0 = 0.6$.

Solution. Program_10b.py is used to plot a pinched hysteresis curve for a HP labs memristor.

```
# Program_10b.py: Pinched hysteresis in the memristor.
import numpy as np
import matplotlib.pyplot as plt
from scipy.integrate import odeint
from matplotlib import style
eta, L, Roff, Ron, p, T, w0 = 1.0, 1.0, 70.0, 1.0, 10.0, 20.0, 0.6
t=np.arange(0.0, 40.0, 0.01)
def memristor(X, t):
    w = X
    dwdt = ((eta * (1 - (2*w - 1) ** (2*p)) * np.sin(2*np.pi * t/T))
            / (Roff - (Roff - Ron) * w))
    return dwdt
X = odeint(memristor, [w0], t, rtol=1e-12)
w = X[:, 0]
style.use("ggplot")
plt.plot(np.sin(2*np.pi * t/T), np.sin(2*np.pi * t/T)
        / (Roff - (Roff - Ron) * X[:, 0]), "g")
plt.xlabel("Voltage", fontsize=15)
plt.ylabel("Current", fontsize=15)
plt.tick_params(labelsize=15)
plt.grid(True)
plt.show()
```

10.2   CHUA'S NONLINEAR ELECTRICAL CIRCUIT

Chua's nonlinear electrical circuit is shown in Figure 10.3(a) and consists of linear RLC components with a nonlinear resistor denoted by N_R. Chua's circuit exhibits many interesting phenomena including period-doubling, intermittency and quasi-periodic routes to chaos [8]. It is simple to build in a laboratory, and connecting the circuit to an oscilloscope gives very interesting results, as shown in Figure 10.3. The differential equations that model the circuit are written in dimensionless form as:

$$\dot{x} = a(y - g(x)), \quad \dot{y} = x - y + z, \quad \dot{z} = -by, \tag{10.3}$$

where a and b are dimensionless parameters. The function $N_R(x)$ has the nonlinear form:

$$N_R(x) = m_1 x + \frac{1}{2}(m_0 - m_1)(|x + 1| - |x - 1|),$$

where a, b, m_0 and m_1, are constants. The parameters x, y, z relate to voltages v_1, v_2 and current i, respectively.

Example 10.2.1. Use Python to produce an animation for the Chua circuit given that $b = 15, m_0 = -\frac{1}{7}, m_1 = \frac{2}{7}$ and $x_0 = 1.96, y_0 = -0.0519, z_0 = -3.077$, as the parameter a increases from $a = 8$ to $a = 11$.

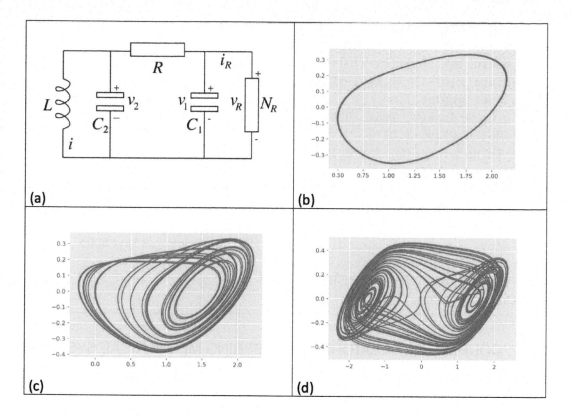

Figure 10.3 Chua's circuit and stills of the animation. (a) The RLC circuit with a nonlinear resistor element N_R. (b) Period one behavior. (c) A single-scroll chaotic attractor. (d) A double-scroll chaotic attractor.

Solution. Program_10c.py solves this system of three ODEs and plots an animation for x versus y, as the parameter a increases from $a = 8$ to $a = 11$. Watching the animation, the reader will see a single periodic solution, period-doubling to a single-scroll chaotic attractor, and then to a double-scroll chaotic attractor. This example provides a wonderful illustration of how mathematics and electrical engineering are closely related. The reader is encouraged to look for videos on YouTube to see how well the model simulates the real world. In 2002, one of my final year project students, Jon Borresen, discovered hysteresis in Chua's circuit for the first time [2].

```
# Program_10c.py: An animation of Chua's circuit in Google Colab.
from matplotlib import pyplot as plt
from matplotlib import animation
from matplotlib import style
import numpy as np
from scipy.integrate import odeint
fig=plt.figure()
myimages=[]
```

```
mo , m1 , Tmax = -1/7 , 2/7 , 50
def Chua(x, t):
    return [a*(x[1]-(m1*x[0]+(mo-m1)/2*(np.abs(x[0]+1)-\
    np.abs(x[0]-1)))),x[0]-x[1]+x[2],-15*x[1]]
style.use('ggplot')
time = np.arange(0, Tmax, 0.1)
x0=[1.96,-0.0519,-3.077]
for a in np.arange(8, 11, 0.1):
    xs = odeint(Chua, x0, time)
    imgplot = plt.plot(xs[:,0], xs[:,1], "g-")
    myimages.append(imgplot)
anim=animation.ArtistAnimation(fig,myimages,interval=500,\
                        blit=False,repeat_delay=500)
plt.close()
from IPython.display import HTML
HTML(anim.to_jshtml())
```

10.3 COUPLED OSCILLATORS: MASS-SPRING MECHANICAL SYSTEMS

Coupled oscillators are important in mechanical engineering, indeed, all solids can be described in terms of coupled oscillations.

Example 10.3.1. Consider a two-mass, three-spring system as depicted in Figure 10.4(a). Using Newton's Second Law of motion, $F = ma$, the equations of motion for the two masses are

$$m_1 \frac{d^2 x_1}{dt^2} = -k_1 x_1 - k_2 (x_1 - x_2), \quad m_2 \frac{d^2 x_2}{dt^2} = -k_2 (x_2 - x_1) - k_3 x_2, \qquad (10.4)$$

where m_1, m_2 are masses, k_1, k_2, k_3 are called spring constants, and $x_1(t), x_2(t)$, are displacements from equilibrium. By writing the velocities, $\dot{x}_1 = y_1$, and $\dot{x}_2 = y_2$, the system of two-second order ODEs (10.4), can instead be written as a system of four first order ODEs

$$\dot{x}_1 = y_1,$$
$$\dot{y}_1 = -\frac{1}{m_1} (k_1 x_1 + k_2 (x_1 - x_2)),$$
$$\dot{x}_2 = y_2,$$
$$\dot{y}_2 = \frac{1}{m_2} (k_2 (x_2 - x_1) + k_3 x_2), \qquad (10.5)$$

where $\dot{y}_1 = \ddot{x}_1$ and $\dot{y}_2 = \ddot{x}_2$. Given that, $m_1 = 1$kg, $m_2 = 2$kg, $k_1 = 1$N per m, $k_2 = 2$N per m, $k_3 = 3$N per m, $x_1(0) = 0$m, $x_2(0) = 1$m, $y_1(0) = 0$ms^{-1} and $y_2(0) = 0$ms^{-1}, in system (10.5), use Python to plot solution curves for $x_1(t)$ and $x_2(t)$.

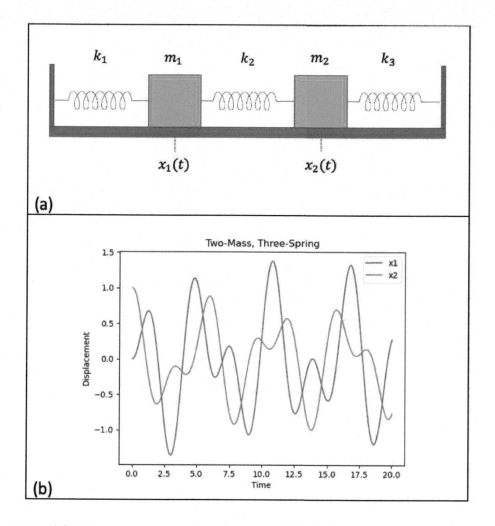

Figure 10.4 (a) A two-mass, three spring system. (b) Displacements of $x_1(t)$ and $x_2(t)$, in metres.

Solution. Program_10d.py solves this system of four ODEs and plots the solution curves for $x_1(t), x_2(t)$, on one graph.

```
# Program 10d.py: Two-mass, three-spring system.
import numpy as np
from scipy.integrate import odeint
import matplotlib.pyplot as plt
# Set parameters and initial conditions
m1 , m2 , k1 , k2 , k3 = 1 , 2 , 1 , 2 , 3
x10, y10, x20, y20 = 0, 0, 1, 0
# Maximum time point and total number of time points
tmax, n = 20, 1001
def Mass_Spring(X, t, m1, m2, k1, k2, k3):
#The Differential Equations
```

```
      x1, y1, x2, y2 = X
      dx1 = y1
      dy1 = -(k1 * x1 + k2 * (x1 - x2)) / m1
      dx2 = y2
      dy2 = -(k2*(x2 - x1) + k3 * x2) / m2
      return (dx1, dy1, dx2, dy2)
# Integrate differential equations on the time grid t.
t = np.linspace(0, tmax, n)
f = odeint(Mass_Spring,(x10, y10, x20, y20),t,args=(m1,m2,k1,k2,k3))
x1, y1, x2, y2 = f.T
plt.figure(1)
plt.xlabel("Time")
plt.ylabel("Displacement")
plt.title("Two-Mass, Three-Spring")
plt.plot(t, x1, label = "x1")
plt.plot(t, x2, label = "x2")
legend = plt.legend(loc = "best")
plt.show()
```

10.4 PERIODICALLY FORCED MECHANICAL SYSTEMS

A periodically forced steel beam between two magnets (or pendulum) can be modeled using the Duffing equation [7] given by

$$\ddot{x} + k\dot{x} + \beta x + \alpha x^3 = \Gamma \cos(\omega t), \qquad (10.6)$$

where, in physical models, x is displacement, $\dot{x}$ represents speed, $k \geq 0$ is the damping coefficient, β is the stiffness, α is the nonlinear stiffness parameter, Γ is the amplitude of force vibration and ω is the frequency of the driving force. When $\beta < 0$, the Duffing equation models a periodically forced steel beam deflected between two magnets [10], see Figure 10.5(a). When $\beta > 0$, the Duffing equation models a periodically forced pendulum, as depicted in Figure 10.5(b). When $\alpha > 0$, the spring is called a hardening spring, and when $\alpha < 0$, the spring is called a softening spring. Let $\dot{x} = y$; then the Duffing equation can be written as a system of the form:

$$\dot{x} = y, \quad \dot{y} = -\beta x - ky - \alpha x^3 + \Gamma \cos(\omega t). \qquad (10.7)$$

Example 10.4.1. Use Python to produce a bifurcation diagram, with feedback, for a periodically forced Duffing system, given that $\alpha = 1$, $\beta = -1$, $k = 0.1$, $\omega = 1.25$, when $0 \leq \Gamma \leq 0.4$. Since $\beta < 0$, this system models the system shown in Figure 10.5(a).

Solution. Program_10e.py shows how to plot a bifurcation diagram with feedback for the Duffing equation (10.7). The amplitude of vibration is ramped up and then ramped back down again. There is a clockwise hysteresis cycle at approximately $0.04 \leq \Gamma \leq 0.08$, there is a bifurcation to period two behavior at approximately

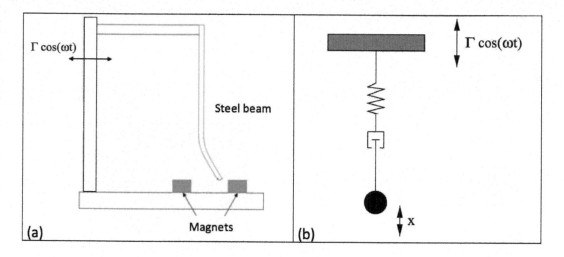

Figure 10.5 (a) A steel beam between two magnets, $\beta < 0$. (b) A periodically forced pendulum, $\beta > 0$.

$0.21 \leq \Gamma \leq 0.23$, and for $\Gamma \geq 0.23$, the system appears to be chaotic with periodic windows. It is important to note in this case that steady-state behavior is a period-one cycle.

```
# Program_10e.py: Bifurcation diagram of the forced Duffing equation.
import matplotlib.pyplot as plt
import numpy as np
from scipy.integrate import odeint
alpha , beta , k , omega = 1 , -1 , 0.1 , 1.25
rs_up , rs_down = [] , []
def Duffing(x, t):
    return [x[1],
            - k*x[1]-beta*x[0]-alpha*x[0]**3+gamma*np.cos(omega*t)]
num_steps , step = 20000 , 0.00002
interval = num_steps * step
x0, y0 = 1, 0
ns = np.linspace(0, num_steps, num_steps)
for n in ns: # Ramp up.
    gamma = step * n
    t = np.linspace(0, (4*np.pi) / omega, 200)
    xs = odeint(Duffing, [x0, y0], t)
    for i in range(2):
        x0 , y0 = xs[100, 0] , xs[100, 1]
        r = np.sqrt(x0**2 + y0**2)
        rs_up.append([n, r])
rs_up = np.array(rs_up)
for n in ns: # Ramp down.
```

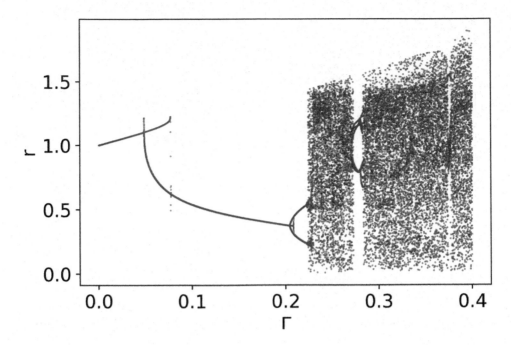

Figure 10.6 Bifurcation diagram of the Duffing equation (10.7) when $\alpha = 1$, $\beta = -1$, $k = 0.1$, $\omega = 1.25$, when $0 \leq \Gamma \leq 0.4$. Note that $r = \sqrt{x^2 + y^2}$. Red points indicate ramp up, $0 \leq \Gamma \leq 0.4$, and blue points represent ramp down, where Γ decreases from $\Gamma = 0.4$ back down to zero. There is a clockwise hysteresis cycle at approximately $0.04 \leq \Gamma \leq 0.08$, and the system is history dependent.

```python
    gamma = interval - step * n
    t = np.linspace(0, (4*np.pi) / omega, 200)
    xs = odeint(Duffing, [x0, y0], t)
    for i in range(2):
        x0 , y0 = xs[100, 0] , xs[100, 1]
        r = np.sqrt(x0**2 + y0**2)
        rs_down.append([num_steps - n, r])
rs_down = np.array(rs_down)
fig, ax = plt.subplots()
xtick_labels = np.linspace(0, interval, 5)
ax.set_xticks([x / interval * num_steps for x in xtick_labels])
ax.set_xticklabels(["{:.1f}".format(xtick) for xtick in xtick_labels])
plt.plot(rs_up[:, 0], rs_up[:,1], "r.", markersize=0.1)
plt.plot(rs_down[:, 0], rs_down[:,1], "b.", markersize=0.1)
plt.xlabel(r"$\Gamma$", fontsize=15)
plt.ylabel("r", fontsize = 15)
plt.tick_params(labelsize = 15)
plt.show()
```

Readers may be interested in our recent publication on coupled oscillators (ions), levitated in a three-dimensional radiofrequency quadrupole ion trap [9].

EXERCISES

10.1 Given that $R = 1\,\Omega$, $L = 1$H, $C = \frac{1}{2}$F, the applied electromotive force (EMF) is $\frac{dE}{dt} = 5e^{-t}$ volts, and $i(0) = \frac{di}{dt}(0) = 0$, in equation (10.1), use Python to solve the ODE and plot a solution curve for the current in the circuit.

10.2 The differential equations used to model two coupled inductor-capacitor (LC) circuits are given by:

$$\frac{d^2 i_1}{dt^2} + \omega^2(1 + \alpha)i_1 + \omega^2 \alpha i_2 = 0, \quad \frac{d^2 i_2}{dt^2} + \omega^2(1 + \alpha)i_2 + \omega^2 \alpha i_1 = 0,$$

where $\omega = \frac{1}{\sqrt{LC}}$, and α is a constant. Given that, $\alpha = 0.2$, and ω is 0.5, $i_1(0) = 1$, $i_2(0) = 0$, $\frac{di_1}{dt}(0) = \frac{di_2}{dt}(0) = 0$, plot solution curves for i_1 and i_2, for $0 \le t \le 100$ seconds.

10.3 Consider a three-mass, four-spring system. The equations of motion for the three masses are

$$m_1 \frac{d^2 x_1}{dt^2} = -k_1 x_1 + k_2 (x_2 - x_1),$$

$$m_2 \frac{d^2 x_2}{dt^2} = -k_2 (x_2 - x_1) + k_3 (x_3 - x_2),$$

$$m_3 \frac{d^2 x_3}{dt^2} = -k_3 (x_3 - x_2) - k_4 x_3,$$

where m_1, m_2, m_3 are masses, k_1, k_2, k_3, k_4 are spring constants and $x_1(t), x_2(t)$, $x_3(t)$, are displacements from equilibrium. Re-write these three second order ODEs as a system of six first order ODEs. Given that, $m_1 = m_2, = m_3 = 1$, $k_1 = 1$, $k_2 = 2$, $k_3 = 3$ and $k_4 = 4$, with $x_1(0) = 1$, $x_2(0) = x_3(0) = 0$ and $\dot{x}_1(0) = \dot{x}_2(0) = \dot{x}_3(0) = 0$, plot solution curves for $0 \le t \le 30$.

10.4 The differential equations used to model the motion of the double pendulum (see Figure 10.7) are given by

$$\ddot{\theta}_1 = \frac{-g(2m_1 + m_2)\sin\theta_1 - m_2 g \sin(\theta_1 - 2\theta_2)}{L_1(2m_1 + m_2 - m_2\cos(2\theta_1 - 2\theta_2))}$$

$$\frac{-2\sin(\theta_1 - \theta_2)m_2\left(\dot{\theta}_2^2 L_2 + \dot{\theta}_1^2 L_1 \cos(\theta_1 - \theta_2)\right)}{L_1(2m_1 + m_2 - m_2\cos(2\theta_1 - 2\theta_2))},$$

$$\ddot{\theta}_2 = \frac{2\sin(\theta_1 - \theta_2)\left(\dot{\theta}_1^2 L_1(m_1 + m_2) + g(m_1 + m_2)\cos\theta_1 + \dot{\theta}_2^2 L_2 m_2 \cos(\theta_1 - \theta_2)\right)}{L_2(2m_1 + m_2 - m_2\cos(2\theta_1 - 2\theta_2))}.$$

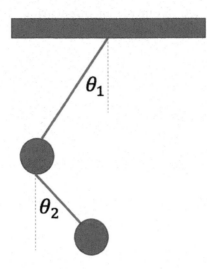

Figure 10.7 The double pendulum.

Use Python to plot phase solutions, θ_1 against θ_2, for the following parameter values:

(i) $g = 9.8, m_1 = m_2 = 1, L_1 = 1.5, L_2 = 1.5, \theta_1(0) = 0.5, \dot{\theta}_1 = 1.5, \theta_2 = 0, \dot{\theta}_2 = 0$;

(ii) $g = 9.8, m_1 = m_2 = 1, L_1 = 6, L_2 = 1, \theta_1(0) = 0.5, \dot{\theta}_1 = 1.5, \theta_2 = 0, \dot{\theta}_2 = 0$.

Vary the initial conditions for $\dot{\theta}_1$ slightly and run the simulations for parts (a) and (b) again. Give a physical interpretation of the results.

Solutions to the Exercises may be found here:

https://drstephenlynch.github.io/webpages/Solutions_Section_2.html.

FURTHER READING

[1] Ashby, D. (2011). *Electrical Engineering 101: Everything You Should Have Learned in School but Probably Didn't*. Newnes, Boston.

[2] Borresen, J. and Lynch, S. (2002). Further investigation of hysteresis in Chua's circuit. *International Journal of Bifurcation and Chaos* **12**(1), 129–134.

[3] Bunnell, B. and Najia, S. (2020). *Mechanical Engineering for Makers: A Hands-on Guide to Designing and Making Physical Things*. Make Community, LLC, Santa Rosa.

[4] Chua, L. (1971). Memristor-missing circuit element. *IEEE Transactions on Circuit Theory.* **CT18**, 507–519.

[5] Hambley, A. (2017). *Electrical Engineering: Principles and Applications.* Pearson, London.

[6] Huber, M. (2013). *The Beginners Guide to Engineering: Mechanical Engineering.* Create Space Independent Publishing Platform. Amazon.com.

[7] Kovacic, I. and Brennan, M.J., eds. (2011), *The Duffing Equation: Nonlinear Oscillators and their Behavior.* Wiley, New York.

[8] Madan, R.N. (1993). *Chua's Circuit: A Paradigm for Chaos.* World Scientific, Singapore.

[9] Mihalcea, B. and Lynch, S. (2021) Investigations on dynamical stability in 3D quadrupole ion traps. *Applied Sciences*, **11**, 2938.

[10] Moon, F.C. and Holmes, P.J. (1979) A magnetoelastic strange attractor, *Journal of Sound and Vibration*, **65**, 276–296.

[11] Strukov, D.B., Snider, G.S., Stewart, D.R., et al. (2008). The missing memristor found. *Nature.* **453**, 80–83.

## CHAPTER 11

Fractals and Multifractals

Fractals were introduced in Section 1.3 of this book using the turtle library. In this chapter, the reader is shown how to plot fractals using matplotlib. The fractal dimension is defined and calculated for simple fractals. The reader is then introduced to the box-counting technique used to compute the fractal dimension of images. Next is an introduction to multifractals, computing their multifractal spectra of dimensions, and looking at applications in the real world. The chapter ends with the Python program for plotting the Mandelbrot set, the most famous multifractal of them all.

Ken Falconer has written a book that provides a quick introduction to fractals [3], he has also written a book considered by many to be the seminal text on fractal geometry [4]. Readers may also be interested in Mandelbrot's book [8], and fractals on the edge of chaos [6]. Fractals are pictorial representations of chaos, and they appear in other chapters of the book. The theory and applications of multifractals are covered in [5].

Aims and Objectives:

- To introduce fractals and fractal dimensions.

- To introduce multifractals by example.

At the end of the chapter, the reader should be able to

- plot simple fractals using matplotlib;

- compute box-counting dimensions of binary images;

- plot multifractal spectra curves;

- plot the Mandelbrot set;

- understand applications in the real world.

11.1 PLOTTING FRACTALS WITH MATPLOTLIB

We start with some simple definitions:

DOI: 10.1201/9781003285816-11

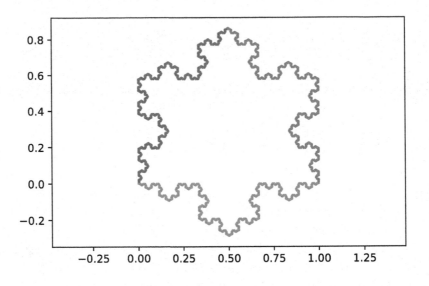

Figure 11.1 The Koch snowflake plotted with matplotlib.

Definition 11.1.1. A fractal is an object that displays self-similarity under magnification and can be constructed using a simple motif.

Definition 11.1.2. The fractal dimension, D_f say, is given as:

$$D_f = \frac{\ln N(\ell)}{-\ln \ell}, \tag{11.1}$$

where N is the number of segments of length ℓ.

Definition 11.1.3. A fractal is an object with non-integer fractal dimension.

Example 11.1.1. Use matplotlib to plot the Koch snowflake and determine the dimension of the fractal object.

Solution. Program_11a.py is used to plot the Koch snowflake, see Figure 11.1. Notice that it has the crude structure of a snowflake in nature. The fractal dimension of the Koch snowflake is

$$D_{\text{Koch}} = \frac{\ln 4}{-\ln \frac{1}{3}} = 1.2619.$$

```
# Program_11a.py: The Koch Snowflake.
import numpy as np
import matplotlib.pyplot as plt
from math import floor
def Koch_Curve(k , angtri , xstart , ystart):
```

```
n_lines , h = 4**k , 3**(-k)
x, y, x[0],y[0] = [0]*(n_lines+1),[0]*(n_lines+1) , xstart , ystart
segment=[0] * n_lines
angle=[0, np.pi/3, -np.pi/3, 0]
for i in range(n_lines):
  m , ang = i , angtri
  for j in range(k):
    segment[j] = np.mod(m, 4)
    m = floor(m / 4)
    ang = ang + angle[segment[j]]
  x[i+1] = x[i]+h*np.cos(ang)
  y[i+1] = y[i]+h*np.sin(ang)
  plt.axis("equal")
  plt.plot(x,y)
k = 6 # Plot the three sides.
Koch_Curve(k , np.pi/3,0,0)
Koch_Curve(k , -np.pi/3 , 0.5 , 0.866)
Koch_Curve(k , np.pi , 1 , 0)
```

Example 11.1.2. Use matplotlib to plot the Sierpinski triangle using the chaos game and determine the dimension of the fractal object.

Solution. The rules of the chaos game are very simple. Consider an equilateral triangle with vertices A, B and C, as depicted in Figure 11.2(a). Start with an initial point x_0 somewhere inside the triangle.

Step 1. Cast an ordinary cubic die with six faces.

Step 2. If the number is either 1 or 2, move half way to the point A and plot a point.

Step 2. Else, if the number is either 3 or 4, move half way to the point B and plot a point.

Step 2. Else, if the number is either 5 or 6, move half way to the point C and plot a point.

Step 3. Starting with the new point generated in Step 2, return to Step 1.

The die is cast again and again to generate a sequence of points $\{x_0, x_1, x_2, x_3, \ldots\}$. As with the other fractals considered here, the mathematical fractal would consist of an infinite number of points. In this way, a chaotic attractor is formed, as depicted in Figure 11.2(b).

Program_11b.py is used to plot the Sierpinski triangle, see Figure 11.2. The fractal dimension of the Sierpinski triangle is

$$D_{\text{Sierp}} = \frac{\ln 3}{-\ln \frac{1}{2}} = 1.5850.$$

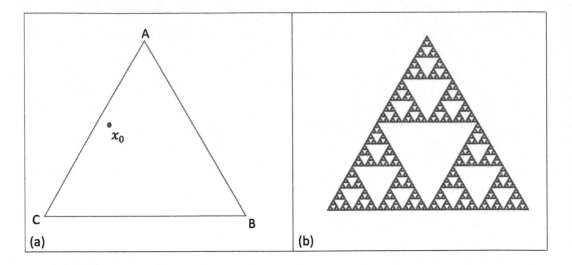

Figure 11.2 (a) The triangle used in the chaos game. (b) The Sierpinski triangle plotted using the chaos game.

```
# Program_11b.py: The Sierpinski triangle and the chaos game.
import matplotlib.pyplot as plt
from random import random, randint
import numpy as np
def midpoint(P, Q):
    return (0.5*(P[0] + Q[0]), 0.5*(P[1] + Q[1]))
# The three vertices A,B,C.
vertices = [(0, 0), (2, 2*np.sqrt(3)), (4, 0)]
iterates = 100000
x, y = [0]*iterates, [0]*iterates
x[0], y[0] = random(), random()
for i in range(1, iterates):
    k = randint(0, 2)
    x[i], y[i] = midpoint(vertices[k], (x[i-1], y[i-1]))
#fig, ax = plt.subplots(figsize=(6, 6))
#ax.scatter(x, y, color='k', s=0.1)
#ax.axis('off')
fig=plt.figure()
plt.scatter(x , y , s = 0.001 , c = "r")
plt.axis("off")
plt.axis("equal")
plt.savefig("Sierpinski.png" , dpi = 400)
plt.show()
```

Example 11.1.3. Use matplotlib to plot the Barnsley fern fractal.

Figure 11.3 The Barnsley fern fractal. The fractal resembles a fern in nature.

Solution. Program_11c.py is used to plot the Barnsley fern fractal, see Figure 11.3. Notice that is has the crude structure of a fern in nature. For details of the transformations required to obtain the fractal, readers are directed to Barnsley's excellent book, Fractals Everywhere [1].

```python
# Program_11c.py: The Barnsley fern fractal.
import matplotlib.pyplot as plt
import random
x , y , N_Iterations = [0], [0], 80000
idx = 0
for i in range(1, N_Iterations):
    r = random.randint(1, 100)
    # Four transformations.
    if r==1:
        x.append(0)
        y.append(0.2 * y[idx]-0.12)
    if r >= 2 and r <= 84:
        x.append(0.845 * x[idx] + 0.035 * y[idx])
        y.append(-0.035 * x[idx] + 0.82 * y[idx] +1.6)
    if r>= 85 and r<= 93:
        x.append(0.2 * x[idx] - 0.31 * y[idx])
        y.append(0.255 * x[idx] + 0.245*(y[idx])+0.29)
    if r >= 93:
        x.append(-0.15 * x[idx] + 0.24 * y[idx])
        y.append(0.25 * x[idx] + 0.2 * y[idx] + 0.68)
```

```
    idx += 1
plt.figure()
plt.scatter(x , y , s = 0.01 , color = "g")
plt.axis("off")
plt.savefig("Fern.png" , dpi = 400)
plt.show()
```

11.2 BOX-COUNTING BINARY IMAGES

The fractal dimension of an image may be computed using the so-called box-counting technique. By covering the object with boxes of varying sizes and counting the number of boxes that contain the object, it is possible to compute the box-counting dimension, which is equivalent to the fractal dimension. The methods used to obtain binary images are covered in the next chapter on image processing.

Example 11.2.1. Use box counting to compute the fractal dimension of the binary image of the Sierpinski triangle displayed in Figure 11.4(a).

Solution. In Program_11d.py, the binary image is loaded as a numpy array and then padded with zeros using **np.pad** to give a 512 × 512 pixel array. Box counting data points are computed and then the **polyfit** function is used to determine the line of best fit. The gradient of the line gives the box-counting dimension of the image, see Figure 11.4(b).

```
# Program 11d.py: Box Counting a Binary Image.
import numpy as np
import matplotlib.pyplot as plt
```

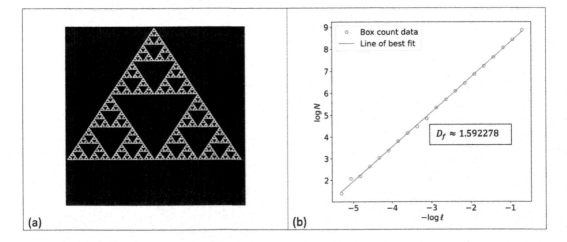

(a)   (b)

Figure 11.4 (a) A padded binary image of the Sierpinski triangle. (b) The line of best fit for the data points. The gradient of the line gives the fractal dimension, $D_f \approx 1.592278$, of the figure displayed in (a).

```python
from skimage import io
import pylab as pl
Sierp_img = io.imread("Sierpinski.png")
plt.imshow(Sierp_img,cmap="gray", interpolation="nearest")
width, height, _ = Sierp_img.shape
binary = np.zeros((width, height))
for i, row in enumerate(Sierp_img):
    for j, pixel in enumerate(row):
        if pixel[0] < 220:
            binary[i, j] = 1
        else:
            binary[i, j] = 0
# Centre the image.
img = binary[70:410,120:540]
width, height = img.shape
# Pad the image.
maxD=max(width,height)
Dim=int(2**np.ceil(np.log2(maxD)))
PadRow = int(Dim-width)
PadCol=int(Dim-height)
image=np.pad(img,[(0,PadRow),(0,PadCol)],mode="constant")
fig=plt.figure()
plt.axis("off")
io.imshow(image)
pixels=[]
for i in range(Dim):
for j in range(Dim):
if image[i,j] ==1:
pixels.append((i,j))
pixels=pl.array(pixels)
scales=np.logspace(1, 8, num=20, endpoint=False, base=2)
Ns=[]
for scale in scales:
    H, edges=np.histogramdd(pixels,bins=(np.arange(0,Dim,scale),\
                            np.arange(0,Dim,scale)))
    Ns.append(np.sum(H > 0))
# Polyfit to a straight line.
coeffs=np.polyfit(np.log(scales), np.log(Ns), 1)
D = -coeffs[0]  # The fractal dimension.
fig=plt.figure()
plt.rcParams["font.size"] = "16"
print("The fractal dimension is ", D )
plt.plot(-np.log(scales),np.log(Ns), "o", mfc="none", \
         label="Box count data")
plt.plot(-np.log(scales), np.polyval(coeffs,np.log(scales)),\
```

```
                label="Line of best fit")
plt.xlabel("$-\log \ell$")
plt.ylabel("$\log N$")
plt.legend()
plt.show()
```

11.3 THE MULTIFRACTAL CANTOR SET

To obtain a multifractal Cantor set, one simply assigns weights (probabilities) to the left and right segments of the fractal as indicated in Figure 11.5(a). The q'th moment (or partition function) Z_q is defined by

$$Z_q = \sum_{i=1}^{N} p_i^q(\ell), \tag{11.2}$$

where p_i are the weights (probabilities) which sum to one, and ℓ is the length scale. For the self-similar multifractal sets, the scaling function $\tau(q)$, satisfies the equation:

$$\sum_{i=1}^{N} p_i^q r_i^{\tau(q)} = 1,$$

with r_i fragmentation ratios. The generalized dimensions, D_q say, and the scaling function $\tau(q)$, are defined by

$$\tau(q) = D_q(1 - q) = \frac{\ln Z_q(\ell)}{-\ln \ell}. \tag{11.3}$$

The $f(\alpha)$ spectrum of dimensions is then determined using the formulae:

$$f(\alpha(q)) = q\alpha(q) + \tau(q), \quad \alpha = -\frac{\partial \tau}{\partial q}. \tag{11.4}$$

Example 11.3.1. Plot the $f(\alpha)$ spectrum of dimensions for the Cantor set, given that $p_1 = \frac{1}{9}$ and $p_2 = \frac{8}{9}$.

Solution. Using equations (11.2) to (11.4), and Program_11.e.py, the $f(\alpha)$ spectrum of dimensions for the Cantor set is shown in Figure 11.5(b).

```
# Program_11e.py: Cantor set multifractal.
from sympy import symbols, log, diff, plot
import matplotlib.pyplot as plt
from sympy.plotting import plot_parametric
plt.rcParams["font.size"] = "16"
p1 , p2 = 1 / 9 , 8 / 9
q , x = symbols("q x")
tau = log(p1**q + p2**q) / log(3)
```

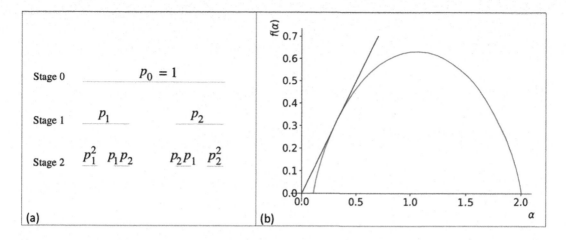

Figure 11.5 (a) A multifractal Cantor set. (b) The $f(\alpha)$ multifractal spectrum when $p_1 = \frac{1}{9}$ and $p_2 = \frac{8}{9}$. The red line is $f(\alpha) = \alpha$.

```
alpha=-diff(tau , q)
f = alpha * q + tau
x = q
p1=plot_parametric(alpha , f ,   (q, -10, 10),xlabel=r'$\alpha$',
                   ylabel=r"$f(\alpha)$",show=False)
p2=plot(x, q, (x, 0, 0.7),line_color="r",show=False)
p1.append(p2[0])
p1.show()
```

Example 11.3.2. Plot the $f(\alpha)$ spectrum of dimensions for the multifractal generated using the motif shown in Figure 11.6(a), given that $p_1 = 0.1$, $p_2 = 0.15$, $p_3 = 0.2$ and $p_4 = 0.55$.

Solution. Using equations (11.2) to (11.4), and Program_11.f.py, the $f(\alpha)$ spectrum of dimensions for the multifractal is shown in Figure 11.6(b).

```
# Program_11f.py: Multifractal f(alpha) spectrum.
from sympy import symbols, log, diff, plot
import matplotlib.pyplot as plt
from sympy.plotting import plot_parametric
plt.rcParams["font.size"] = "16"
p1 , p2 , p3 , p4 = 0.1 , 0.15 , 0.2 , 0.55
q , x = symbols("q x")
tau = log(p1**q + p2**q + p3**q + p4**q) / log(2)
alpha=-diff(tau , q)
f = alpha * q + tau
x = q
p1=plot_parametric(alpha , f ,   (q, -15, 10),xlabel=r'$\alpha$',
```

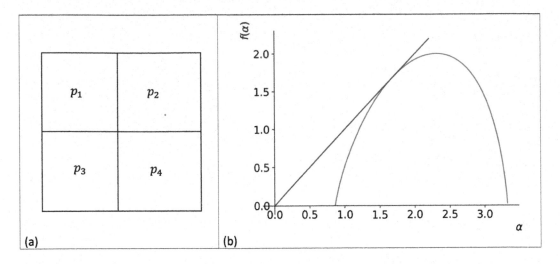

Figure 11.6 (a) A multifractal set planar motif. (b) The $f(\alpha)$ multifractal spectrum when $p_1 = 0.1$, $p_2 = 0.15$, $p_3 = 0.2$ and $p_4 = 0.55$. The red line is $f(\alpha) = \alpha$.

```
                    ylabel=r"$f(\alpha)$",show=False)
p2=plot(x, q, (x, 0, 2.2),line_color="r",show=False)
p1.append(p2[0])
p1.show()
```

Multifractals are very useful in applications when measuring density, dispersion and clustering of objects on surfaces. There are a wide range of applications, for example, see [9] for predicting flame retardant filler/polymer composites in materials science, [10] for investigating electrochemical deposition in the manufacture of graphene macro electrodes, [11] for identifying how surface roughness affects fungal spore binding, and [2] and [12] for quantifying the pattern of microbial cell dispersion of surfaces with different chemistries and topographies.

11.4   THE MANDELBROT SET

The Mandelbrot set is the set of complex numbers, c say, which remain bounded on iteration of the function:

$$f_c(z_n) = z_{n+1} = z_n^2 + c,$$

from $z_0 = 0$. Pixels are colored according to how quickly the iterates cross a certain boundary. For more details, the interested reader is directed to [7].

Example 11.4.1. Plot the Mandelbrot set with Python.

Solution. Program_11g.py lists the code to plot the Mandelbrot set displayed in Figure 11.7. The reader is encouraged to look for videos on the web of a Mandelbrot set zoom.

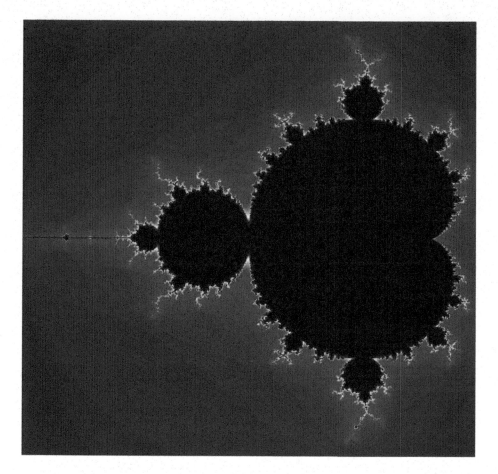

Figure 11.7 The Mandelbrot set. The object is a mathematical constant, much like the number π, and it contains an infinite amount of detail.

```python
# Program_11g.py: The Mandelbrot Set.
import matplotlib.pyplot as plt
import numpy as np
import matplotlib.cm as cm
xmin, xmax, ymin, ymax = -2.5, 1, -1.5, 1.5
xrange = xmax - xmin
yrange = ymax - ymin
def get_iter(c:complex, thresh:int =4, max_steps:int =25) -> int:
    z, i = c, 1
    while i<max_steps and (z*z.conjugate()).real<thresh:
        z = z * z + c
        i+=1
    return i
def plotter(n, thresh, max_steps = 25):
    mapper = lambda x, y: (4*(x-n//2)/n, 4*(y-n//2)/n)
    img = np.full((n,n), 255)
```

```
        x_lower = 0
        x_upper = 5*n//8
        y_lower = 2*n//10
        y_upper = 8*n//10
        for x in range(x_lower, x_upper):
            for y in range(y_lower, y_upper):
                it = get_iter(complex(*mapper(x,y)), thresh=thresh, \
                max_steps=max_steps)
                img[y][x] = 255 - it
        return img[y_lower:y_upper, x_lower:x_upper]
n=1000
img = plotter(n, thresh=4, max_steps=50)
fig, ax = plt.subplots(figsize=(7,7))
plt.imshow(img, cmap = cm.seismic)
plt.axis("off")
plt.show()
```

EXERCISES

11.1 Edit Program_11a.py to plot a Koch square fractal, where Koch square curves are joined to the outer edges of a unit square. Determine the area bounded by the true fractal.

11.2 Use box-counting to estimate the fractal dimension of Barnsley's fern.

11.3 Plot a multifractal $f(\alpha)$ spectrum for the Koch curve multifractal with motif shown in Figure 11.8, given that $p_1 = \frac{1}{9}$ and $p_2 = \frac{1}{3}$.

11.4 Plot a multifractal $f(\alpha)$ spectrum for the plane with motif matrix:

$$M = \begin{pmatrix} 0.05 & 0.1 & 0.2 \\ 0.04 & 0.3 & 0.03 \\ 0.15 & 0.06 & 0.07 \end{pmatrix}.$$

In Chapter 12, you will be asked to obtain an image of this multifractal using image processing techniques.

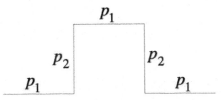

Figure 11.8 Motif for a Koch curve multifractal.

Solutions to the Exercises may be found here:

`https://drstephenlynch.github.io/webpages/Solutions_Section_2.html.`

FURTHER READING

[1] Barnsley, M.F. (2012). *Fractals Everywhere, 3rd Ed.* Dover Publications, New York.

[2] Evans, A., Slate, A.J., Tobin, M. et al. (2022). Multifractal analysis to determine the effect of surface topography on the distribution, density, dispersion and clustering of differently organised coccal shaped bacteria. *Antibiotics.* **11**, 11050551.

[3] Falconer, K. (2003). *Fractals: A Very Short Introduction.* Oxford University Press, Oxford.

[4] Falconer, K. (2014). *Fractal Geometry: Mathematical Foundations and Applications.* Wiley, New York.

[5] Harte, D. (2001). *Multifractals: Theory and Applications.* Chapman and Hall, London.

[6] Linton, O. (2021). *Fractals: On the Edge of Chaos.* Bloomsbury Publishing, London.

[7] Mandelbrot, B.B. (2014). *Fractals and Chaos: The Mandelbrot Set and Beyond.* Springer, New York.

[8] Mandelbrot, B.B. (2020). *Fractals: Form, Chance and Dimension.* Echo Point Books and Media, Reprint ed.

[9] Mills, S.L., Lees G.C., Liauw C.M. et al. (2005). Prediction of physical properties following the dispersion assessment of flame retardant filler-polymer composites based on the multifractal analysis of SEM images. *J. Macromolecular Sci. B- Physics.* **44**(6), 1137–1151.

[10] Slate, A.J., Whitehead, K.A., Lynch, S. et al. (2020). Electrochemical decoration of additively manufactured graphene macro electrodes with MoO2 nanowire: An approach to demonstrate the surface morphology. *J. of Physical Chemistry C.* **124**(28), 15377–15385.

[11] Whitehead, K.A., El Mohtadi M., Lynch, S. et al. (2021). Diverse surface properties reveal that substratum roughness affects fungal spore binding. *iScience.* **24**(4), 102333.

[12] Wickens, D., Lynch, S., Kelly, P. et al. (2014). Quantifying the pattern of microbial cell dispersion, density and clustering on surfaces of differing chemistries and topographies using multifractal analysis. *J. of Microbiological Methods.* **104**, 101–108.

## CHAPTER 12

Image Processing

Image Processing is a branch of Computer Science and is the process of transforming an image into a digital form to perform certain operations to get useful information from it. The first example is concerned with grayscale images, the reader is shown how to create an image of a multifractal. The second example is concerned with a color image and determining the number of pixels of a certain color. In the third example, an image of microbes on a surface is analyzed using statistical methods. The chapter ends with two examples from medical imaging. In the first example, an image of the retinal vascular architecture in a human retina is obtained using filtering. Optometrists can monitor these images for potential retinal, systemic, neurological and cerebral vascular morbidities and risks. In the last example, segmentation and masking is used to detect a tumor in a human brain.

There is extensive online documentation accompanying image processing and analysis with Python. The author has used scikit-image in this book:

`https://scikit-image.org`.

Scikit-image is a collection of algorithms for image processing and is packaged with Anaconda. It includes algorithms for analysis, color manipulation, contrast and exposure, cropping, feature detection, filtering, geometric transformations, morphology, segmentation and more. It is designed to interoperate with the Python numerical and scientific libraries numpy and scipy.

It is recommended that the reader read through the scikit-image web pages before continuing with this chapter. Other packages are available, for example, OpenCV and Python Image Library (Pillow/PIL). There are many books published on image processing, mainstream books include [2] and [3], and [4] and [5] are specialist Python books. For those interested in bio-signal and medical image processing, see [1] and [6].

Aims and Objectives:

- To provide a tutorial guide to image processing.

- To show how to manipulate images.

- To provide tools to analyze images.

DOI: 10.1201/9781003285816-12

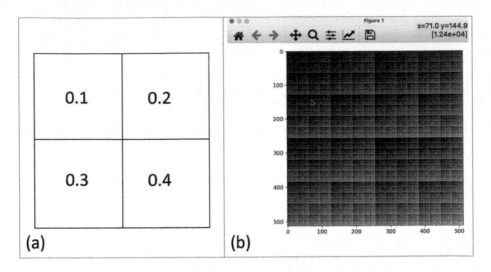

Figure 12.1 (a) A weight distribution motif. (b) The multifractal image. To get Figure 1, in Spyder, under **Preferences**, make sure **Graphics backend** is set to **Automatic**. The value of the pixel at (71.0,144.9) is 1.24e+04, as indicated in the top right hand corner of the figure.

At the end of the chapter, the reader should be able to

- load and save images;

- perform analysis on color, grayscale, and black and white images;

- perform simple image processing on medical images.

12.1 IMAGE PROCESSING, ARRAYS AND MATRICES

Example 12.1.1. Using Spyder, construct an image of a grayscale multifractal lattice using a given matrix motif. A multifractal is a heterogeneous fractal.

Solution. Program_12a.py shows how to construct a 512×512 matrix of pixel values by applying a simple motif to smaller and smaller squares on a lattice. Figure 12.1 shows the motif and the grayscale image after adjusting its contrast. As you scroll across the figure with the mouse, you will notice that the numbers in the top right corner of the image change. Notice that the x values are on the x-axis and read from left to right. The y-values run from top to bottom on the vertical axis in image analysis. The value of the pixel at coordinate position (71.0,144.9) is equal 1.24e+04, as displayed in Figure 12.1(b).

```
# Program_12a.py: Generating a multifractal image.
import numpy as np
import matplotlib.pyplot as plt
from skimage import exposure, io, img_as_ubyte
```

```
p1, p2, p3, p4 = 0.1, 0.2, 0.3, 0.4          # See motif.
p = [[p1, p2], [p3, p4]]
for k in range(1, 9, 1):                      # 512x512 image.
    M = np.zeros([2 ** (k + 1), 2 ** (k + 1)])
    M.tolist()
    for i in range(2**k):
        for j in range(2**k):
            M[i][j] = p1 * p[i][j]
            M[i][j + 2**k] = p2 * p[i][j]
            M[i + 2**k][j] = p3 * p[i][j]
            M[i + 2**k][j + 2**k] = p4 * p[i][j]
    p = M
M = exposure.adjust_gamma(M, 0.2)             # Change the contrast.
plt.imshow(M, cmap="gray", interpolation="nearest")
im = np.array(M, dtype="float64")
im = exposure.rescale_intensity(im, out_range="float")
im = img_as_ubyte(im)
# Save the image as a portable network graphics (png) image.
io.imsave("Multifractal.png", im)
io.imshow(im)
```

12.2 COLOR IMAGES

Example 12.2.1. Determine the number of red pixels in a color image of peppers.

Solution. Program 12b.py shows how to count colored pixels in an image. The image **peppers.jpeg** (see Figure 12.2(a)) has to be in the same folder as **Program_12b.py**. The command **peppers.shape** gives the dimensions of the image as $(700, 700, 3)$, which means that the image is 700×700 pixels, and each pixel has an RGB value associated with it. The next command, **peppers[400,400]**, lists the RGB values of the pixel at coordinate position (400,400), as the array [226 2 12], and the data type is uint8, which means that the RGB values each range from 0 to 255. Program_12b.py takes pixels with red values greater than 190, green values less than 120, and blue values less than 170, and counts these as red pixels. These pixels are then assigned to the value one, the other pixels are assigned the value zero, leading to the binarized image in Figure 12.2(b). Finally, the **sum** function is used to determine the total number of white pixels (ones), and hence the number of red pixels in the original image.

```
# Program_12b.py: Counting colored pixels.
from skimage import io
import numpy as np
import matplotlib.pyplot as plt
peppers = io.imread("peppers.jpeg")
plt.figure(1)
```

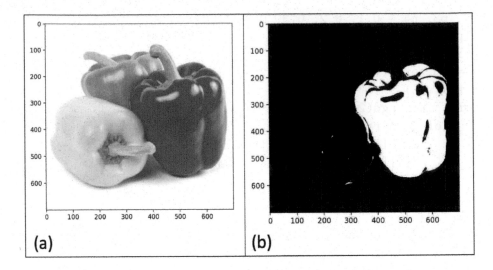

Figure 12.2 (a) Color image of peppers. (b) Binarized image of the red pixels. There are approximately 96,897 red pixels in the original image **peppers.jpeg**. Notice that some red pixels are missed due to lighting.

```
io.imshow(peppers)
print("Image Dimensions=" , peppers.shape)
print("peppers[100,100]=",peppers[400,400])
Red = np.zeros((700,700))
for i in range(700):
    for j in range(700):
        if peppers[j,i,0]>190 and peppers[j,i,1]<120 \
            and peppers[j,i,2]<170:
            Red[j,i]=1
        else:
            Red[j,i]=0
plt.figure(2)
plt.imshow(Red,cmap="gray")
pixel_count = int(np.sum(Red))
print("There are {:,} red pixels".format(pixel_count))
```

12.3   STATISTICAL ANALYSIS ON AN IMAGE

Example 12.3.1. Perform some statistical analysis on the image of microbes on a surface (obtained using a scanning electron microscope).

Solution. See Figure 12.3. Program_12c.py reads in the image **microbes.png**, converts the grayscale image into a binary image, determines the coordinates of the centroids of the clusters, produces a histogram of the areas of the clusters against the number of clusters, and finally, uses the Canny edge detector to determine the edges of clusters

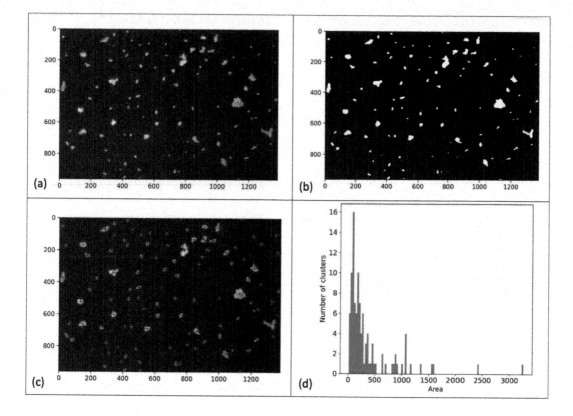

Figure 12.3 (a) The original image, microbes.png. (b) A binary image of the clusters of microbes. (c) The centroids of the clusters overlaying the original image. (d) Histogram of the data.

of microbes. The image is from our research paper on the effect of surface properties on bacterial retention [7], in which we use multifractal analysis to give quantitative measures of density, dispersion and clustering. Multifractals are discussed in Chapter 11.

There are 106 clusters of cells and a total of 38,895 white pixels. The program also prints out the individual areas of the clusters. Note that the program also plots the edges of the clusters (fig5 in the program) but the picture would not be clear in the book.

```
# Program_12c.py: Statistical Analysis on Microbes.
import matplotlib.pyplot as plt
from skimage import io , measure
import numpy as np
from skimage.measure import regionprops
from scipy import ndimage
from skimage import feature
microbes_img = io.imread("Microbes.png")
fig1 = plt.figure() # Original image.
```

```
plt.imshow(microbes_img,cmap="gray", interpolation="nearest")
width, height, _ = microbes_img.shape
fig2 = plt.figure() # Binary image.
binary = np.zeros((width, height))
for i, row in enumerate(microbes_img):
    for j, pixel in enumerate(row):
        if pixel[0] > 80:
            binary[i, j] = 1
plt.imshow(binary,cmap="gray")
print("There are {:,} white pixels".format(int(np.sum(binary))))
blobs = np.where(binary>0.5, 1, 0)
labels, no_objects = ndimage.label(blobs)
props = regionprops(blobs)
print("There are {:,} clusters of cells:".format(no_objects))
# fig3. Centroids of the clusters.
object_labels = measure.label(binary)
some_props=measure.regionprops(object_labels)
fig,ax = plt.subplots(1,1)
#plt.axis('off')
ax.imshow(microbes_img,cmap="gray")
centroids = np.zeros(shape=(len(np.unique(labels)),2))
for i , prop in enumerate(some_props):
    my_centroid = prop.centroid
    centroids[i,:]=my_centroid
    ax.plot(my_centroid[1],my_centroid[0],"r+")
#print(centroids)
fig4 = plt.figure() # Histogram of the data.
labeled_areas = np.bincount(labels.ravel())[1:]
print(labeled_areas)
plt.hist(labeled_areas,bins=no_objects)
plt.xlabel("Area",fontsize=15)
plt.ylabel("Number of clusters",fontsize=15)
plt.tick_params(labelsize=15)
fig5 = plt.figure() # Canny edge detector.
edges=feature.canny(binary,sigma=2,low_threshold=0.5)
plt.imshow(edges,cmap=plt.cm.gray)
plt.show()
```

12.4 IMAGE PROCESSING ON MEDICAL IMAGES

Medical image processing usually encompasses the use and exploration of 3D image datasets of the human body obtained using a Magnetic Resonance Imaging (MRI) scanner or a Computed Tomography (CT) scanner, that uses X-rays and a computer to create detailed images of the inside of the body. This section will consider 2D images only.

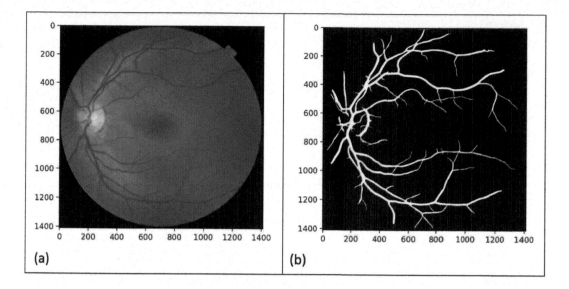

Figure 12.4 (a) The original image of a human retina. (b) Vascular architecture tracing using the **sato** ridge filter.

Example 12.4.1. Obtain an image of the retinal vascular architecture from the image of a human retina.

Solution. See Figure 12.4. Program_12d.py reads in the image of a human retina (see Figure 12.4(a)) from the skimage data set, it then converts the image to grayscale and a **sato** filter is applied to obtain a binary image of the vascular architecture in the retina (see Figure 12.4(b)).

```
# Program_12d.py: Vascular architecture tracing using ridge filters.
import matplotlib.pyplot as plt
from skimage.color import rgb2gray
from skimage import color, data, filters
from skimage import morphology
retina_source = data.retina()
_, ax = plt.subplots()
ax.imshow(retina_source)
retina = color.rgb2gray(retina_source)
t0, t1 = filters.threshold_multiotsu(retina, classes=3)
mask = (retina > t0)
vessels = filters.sato(retina, sigmas=range(1, 10)) * mask
img_gray = rgb2gray(vessels)
t = 0.015
binary_mask = img_gray > t
fig, ax = plt.subplots()
```

```
# Remove small objects.
binary_mask = morphology.remove_small_objects(binary_mask, 600)
plt.imshow(binary_mask, cmap="gray")
plt.show()
```

Example 12.4.2. Use image segmentation and masking to identify tumors and other features in an image of a human brain.

Solution. See Figure 12.5. Image segmentation is the process of partitioning an image into multiple segments. A Simpler Linear Iterative Clustering (SLIC) algorithm is used to generate superpixels, which are groups of pixels that share common characteristics, like pixel intensity. Masking is used to reduce the image noise and improve segmentation results.

```
# Program_12e.py: Segmentation of a human brain.
import matplotlib.pyplot as plt
from skimage import color
from skimage import morphology
from skimage import segmentation
import matplotlib.image as img
num_segs , threshold = 100 , 0.82
img =img.imread("Brain.jpg")
# Compute a mask
lum = color.rgb2gray(img)
mask = morphology.remove_small_holes(
    morphology.remove_small_objects(lum > threshold , 500) , 500)
mask = morphology.opening(mask, morphology.disk(3))
# SLIC result
slic = segmentation.slic(img, n_segments=num_segs, start_label=1)
# SLIC result in Mask
m_slic = segmentation.slic(img, n_segments=num_segs, mask=mask, \
start_label=1)
# Display result
fig, ax_arr = plt.subplots(2, 2, sharex=True, sharey=True, \
figsize=(10, 10))
ax1, ax2, ax3, ax4 = ax_arr.ravel()
ax1.imshow(img)
ax1.set_title("Brain.jpg")
ax2.imshow(mask, cmap="gray")
ax2.set_title("Mask")
ax3.imshow(segmentation.mark_boundaries(img, slic))
ax3.contour(mask, colors="red", linewidths=1)
ax3.set_title("SLIC")
ax4.imshow(segmentation.mark_boundaries(img, m_slic))
ax4.contour(mask, colors="red", linewidths=1)
```

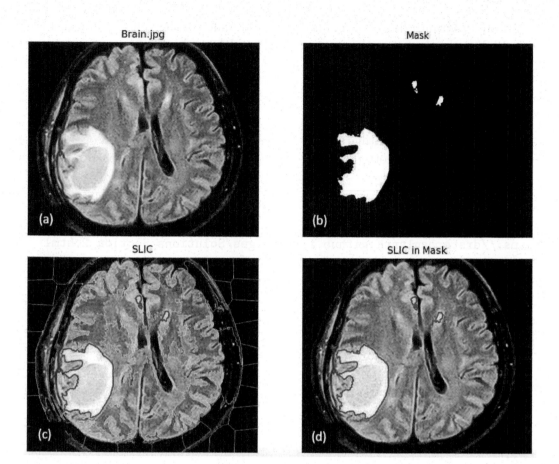

Figure 12.5 (a) The original image of a human brain with a tumor. (b) Compute a mask to identify the tumor. (c) SLIC image. (d) SLIC in the Mask.

```
ax4.set_title("SLIC in Mask")
for ax in ax_arr.ravel():
    ax.set_axis_off()
plt.tight_layout()
plt.show()
```

EXERCISES

12.1 Use the matrix motif:

$$M = \begin{pmatrix} 0.05 & 0.1 & 0.2 \\ 0.04 & 0.3 & 0.03 \\ 0.15 & 0.06 & 0.07 \end{pmatrix},$$

to obtain a 729×729 pixel multifractal image.

12.2 Edit Program_12b.py to obtain a binarized image of the yellow pixels in Figure 12.2(a). Look up on the internet how to delete objects from an image, and determine the number of yellow pixels in the original image.

12.3 Determine an approximation for the length of the vessels of the image in Figure 12.4(b).

12.4 Determine the percentage area of the tumor compared to the area of the human brain (Brain.jpg) plotted in Figure 12.5(a). Do not include any part of the skull. Plot images to represent your results.

Solutions to the Exercises may be found here:

https://drstephenlynch.github.io/webpages/Solutions_Section_2.html.

FURTHER READING

[1] Birkfellener, M. (2017). *Applied Medical Image Processing, 2nd Ed: A Basic Course.* Taylor & Francis, New York.

[2] Burger, W. and Burge M.J. (2013). *Principles of Digital Image Processing: Advanced Methods.* Springer, New York.

[3] Gonzalez, R.C. (2018). *Digital Image Processing, 4th Ed.* Pearson, India.

[4] Dey, S. (2020). *Python Image Processing Cookbook: Over 60 recipes to help you perform complex image processing and computer tasks with ease.* Packt Publishing, Birmingham, UK.

[5] Pajankar, A. (2019). *Python 3 Image Processing: Learn Image Processing with Python 3, Numpy, Matplotlib, and SciKit-image.* BPB Publications, India.

[6] Stemlow, J.L. and Griffel, B. (2014). *Biosignal and Medical Image Processing 3rd Ed.* CRC Press, New York.

[7] Tetlow, L., Lynch, S. and Whitehead, K. (2017). The effect of surface properties on bacterial retention: a study utilising stainless steel and TiN/25.65 at %Ag substrata. *Food and Bioproducts Processing* **102**, 332–339.

## CHAPTER 13

Numerical Methods for Ordinary and Partial Differential Equations

Numerical methods is a branch of mathematics that uses numerical approximation for problems in mathematical analysis. There are many books published on numerical methods, readers will find the following Python books of interest [3], [4] and [5]. Since this book is concerned with scientific computing and modeling real-world problems, the focus is on the numerical solution of ordinary differential equations (ODEs) and initial value problems (IVPs), and partial differential equations (PDEs) and boundary value problems (BVPs). The first section is concerned with Euler's method which uses linear approximation and the Taylor series, taking small tangent lines over short distances, to approximate a solution to an initial value problem. Euler's method is a first order, single step method which fails for certain systems. A more robust, and arguably most useful, multi-step solver is the fourth-order Runge-Kutta method [2], also known as RK4. The second section lists the algorithm for this solver and a simple example is presented. The third and fourth sections are concerned with PDEs, involving functions of several variables, which have a huge range of applications throughout science. Finite difference methods are used in the third and fourth sections when modeling heat in an insulated rod and the vibration of a string, respectively. More examples are covered in the exercises at the end of the chapter. For the theory of PDEs, readers are directed to [1] and [6], for example.

Readers should note that Python has several built-in functions for solving initial value problems, including **solve_ivp** and **odeint**, from the **SciPy** library.

Aims and Objectives:

- To introduce the Euler and Runge Kutta methods for ODEs.

- To introduce numerical methods for PDEs.

At the end of the chapter, the reader should be able to

DOI: 10.1201/9781003285816-13

- solve ODEs numerically using Euler and Runge Kutta methods;

- use finite difference approximations to solve certain PDEs numerically.

13.1 EULER'S METHOD TO SOLVE IVPS

Euler's method is the simplest first-order numerical scheme for solving IVPs of the form:

$$\frac{dy}{dx} = f(x, y(x)), \quad y(x_0) = y_0. \tag{13.1}$$

It is first-order as the local error is proportional to the square of the step-size, h say, and the global error is proportional to h. Euler's method may be derived using a geometrical approach as depicted in Figure 13.1(a), or by considering the Taylor series expansion of the function

$$y(x_n + h) = y(x_n) + h\frac{dy}{dx}(x_n) + \frac{1}{2}h^2\frac{d^2y}{dx^2}(x_n) + \mathcal{O}\left(h^3\right), \tag{13.2}$$

where h is the step-size. Ignoring the quadratic and higher order terms in equation (13.2), and using the first part of equation (13.1), we have

$$y(x_n + h) = y(x_{n+1}) = y(x_n) + hf(x_n, y_n).$$

Writing $y(x_{n+1}) = y_{n+1}$ and $y(x_n) = y_n$, leads to Euler's iterative formula.

Definition 13.1.1. Euler's explicit iterative formula to solve an IVP of the form (13.1), is defined by

$$y_{n+1} = y_n + hf(x_n, y_n). \tag{13.3}$$

This iterative formula is easily implemented in Python.

Example 13.1.1. Use Euler's method to solve the IVP:

$$\frac{dy}{dx} = 1 - y(x), \quad y(0) = 2. \tag{13.4}$$

Solution. Program_13a.py applies Euler's method to the IVP. Note that, in this case, the analytical solution can be computed with Python and is equal to $y(x) = 1 + e^{-x}$. A comparison of the numerical and analytical solution curves is shown in Figure 13.1(b).

```
# Program_13a.py: Eulers Method for an IVP.
import numpy as np
import matplotlib.pyplot as plt
f = lambda x, y: 1 - y      # The ODE.
h , y0 = 0.1 , 2            # Step size and y(0).
x = np.arange(0, 1 + h, h) # Numerical grid.
y , y[0] = np.zeros(len(x)) , y0
```

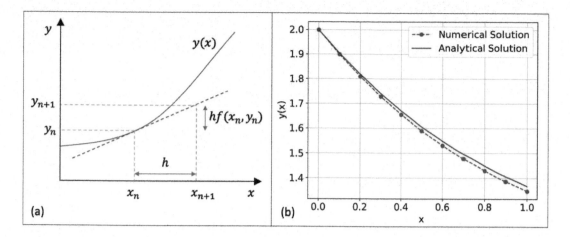

Figure 13.1 (a) Geometrical approach to Euler's method. The red dashed line is a tangent to the curve $y(x)$ at the point (x_n, y_n). See equation (13.3). (b) Numerical and analytical solution of the IVP (13.4) when $h = 0.1$. The numerical solution may be improved by decreasing the step-size h.

```
for n in range(0, len(x) - 1):
    y[n + 1] = y[n] + h*f(x[n] , y[n])
plt.rcParams["font.size"] = "16"
plt.figure()
plt.plot(x, y, "ro--", label='Numerical Solution')
plt.plot(x, np.exp(-x) + 1, "b", label="Analytical Solution")
plt.xlabel("x")
plt.ylabel("y(x)")
plt.grid()
plt.legend(loc="upper right")
plt.show()
```

13.2 RUNGE KUTTA METHOD (RK4)

One can increase the accuracy of the Euler method by decreasing the step size, however, this is not efficient, and may not be stable. As an alternative, using a higher-order method can increase efficiency and accuracy. The most popular higher-order method is the fourth order Runge-Kutta (RK4) method that takes a sampling of four slopes, k_1 to k_4 say, in an interval, and takes an average to approximate y_{n+1}. See Figure 13.2(a) for a geometrical interpretation. It is fourth-order as the local truncation error is on the order of $\mathcal{O}\left(h^5\right)$, and the global error is $\mathcal{O}\left(h^4\right)$.

Definition 13.2.1. The iterative formula for the fourth-order Runge Kutta method is

$$
\begin{aligned}
y_{n+1} &= y_n + \frac{1}{6}\left(k_1 + 2k_2 + 2k_3 + k_4\right), \\
k_1 &= hf\left(x_n, y_n\right), \\
k_2 &= hf\left(x_n + \frac{h}{2}, y_n + \frac{k_1}{2}\right), \\
k_3 &= hf\left(x_n + \frac{h}{2}, y_n + \frac{k_2}{2}\right), \\
k_4 &= hf\left(x_n + h, y_n + k_3\right).
\end{aligned}
\tag{13.5}
$$

The algorithm is easily implemented in Python, compare Program_13a.py and Program_13b.py.

Example 13.2.1. Use the RK4 algorithm, equations (13.5), to solve the IVP (13.4).

Solution. Program 13b.py lists the RK4 method for solving the IVP (13.4), Figure 13.2(b) shows the numerical and analytical solutions, which are indistinguishable in this case.

```
# Program_13b.py: RK4 Method for an IVP.
import numpy as np
import matplotlib.pyplot as plt
f = lambda x, y: 1 - y      # The ODE.
h , y0 = 0.1 , 2            # Step size and y(0).
x = np.arange(0, 1 + h, h) # Numerical grid.
y , y[0] = np.zeros(len(x)) , y0
for n in range(0, len(x) - 1):
```

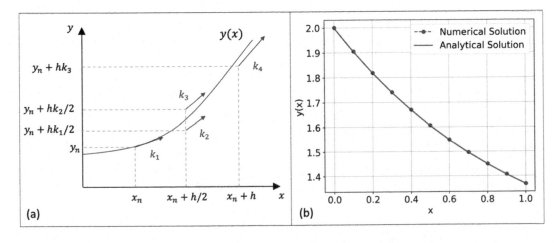

Figure 13.2 (a) Geometrical interpretation of the RK4 method. (b) Numerical and analytical solution of the IVP (13.4) using RK4.

```
    k1=h*f(x[n] , y[n])
    k2=h*f(x[n] + h/2 , y[n] + k1/2)
    k3=h*f(x[n] + h/2 , y[n] + k2/2)
    k4=h*f(x[n] + h , y[n] + k3)
    k=(k1 + 2 * k2 + 2 * k3 + k4) / 6
    y[n+1] = y[n] + k
plt.rcParams["font.size"] = "16"
plt.figure()
plt.plot(x, y, "ro--", label="Numerical Solution")
plt.plot(x, np.exp(-x) + 1, "b", label="Analytical Solution")
plt.xlabel("x")
plt.ylabel("y(x)")
plt.grid()
plt.legend(loc="upper right")
plt.show()
```

There are many numerical methods available, and implementation of the iterative formulae is straightforward with Python. The reader will be asked to carry out a literature survey on the backward (implicit) Euler method and the two-step implicit Adams-Bashforth method in order to solve problems in the exercises at the end of the chapter.

13.3 FINITE DIFFERENCE METHOD: THE HEAT EQUATION

The remaining sections provide an introduction to the finite difference method (FDM) for solving PDEs. Other numerical methods are available and the interested reader can research stability, boundary conditions, grid independence, verification and validation, see [1] and [6], for example. There are three famous PDEs, namely:

The 1D Diffusion Equation: This is a parabolic PDE and can be written as

$$\frac{\partial U}{\partial t} = \alpha \frac{\partial^2 U}{\partial x^2}, \tag{13.6}$$

and can be used to model the heat in an insulated rod, see Figure 13.3(a).

The Wave Equation: This is a hyperbolic PDE and can be written as

$$\frac{\partial^2 U}{\partial t^2} = c^2 \frac{\partial^2 U}{\partial x^2}, \tag{13.7}$$

and can be used to model a vibrating guitar string.

Laplace's Equation: This is an elliptic PDE and can be written as

$$\frac{\partial^2 U}{\partial x^2} + \frac{\partial^2 U}{\partial y^2} = 0, \tag{13.8}$$

and can be used to model steady-state temperature distribution in a plate.

In order to solve these, and other PDEs, one may construct an FDM toolkit from Taylor series expansions in the following way. Now for the spatial variable x:

$$U\left(t, x_i + \Delta x\right) = U\left(t, x_i\right) + \Delta x U_x\left(t, x_i\right) + \frac{\Delta x^2}{2!} U_{xx}\left(t, x_i\right) + \mathcal{O}\left(\Delta x^n\right),$$

where $U_x = \frac{\partial U}{\partial x}$ and $U_{xx} = \frac{\partial^2 U}{\partial x^2}$. Suppose that $t = t_n$, taking the first two terms on the right and rearranging the equation gives

$$U_x\left(t_n, x_i\right) = \frac{U\left(t_n, x_{i+1}\right) - U\left(t_n, x_i\right)}{\Delta x},$$

where $x_{i+1} = x_i + \Delta x$, and introducing the notation:

$$U_i^n = U\left(t_n, x_i\right),$$

leads to the first order forward finite difference approximation (FDA):

$$U_x \approx \frac{U_{i+1}^n - U_i^n}{\Delta x}.$$

The other FDAs in Table 13.1, may be derived in a similar fashion and can be used to create finite difference schemes to obtain approximate solutions to many PDEs.

Example 13.3.1. The heat diffusion along an insulated metal bar of length, L say, can be modeled with the PDE

$$U_t - \alpha U_{xx} = 0, \tag{13.9}$$

where α is thermal diffusivity, given by

$$\alpha = \frac{k}{\rho c_p},$$

where k is thermal conductivity, ρ is density and c_p is specific heat capacity. Given that $L = 1$, $\alpha = 0.1$. $U(t, 0) = 0$, $U(t, L) = 0$, and the initial temperature in the rod is $U(0, x) = \sin\left(\frac{\pi x}{L}\right)$, use suitable FDAs and Python to plot solution curves for $0 \le t \le 1$. From Table 13.1, using a forward difference in time FDA and a symmetric difference in space FDA, equation (13.9) becomes:

$$\frac{U_i^{n+1} - U_i^n}{\Delta t} - \alpha \left(\frac{U_{i+1}^n - 2U_i^n - U_{i-1}^n}{\Delta x^2}\right) = 0,$$

and rearranging:

$$U_i^{n+1} = U_i^n + \frac{\alpha \Delta t}{\Delta x^2} \left(U_{i+1}^n - 2U_i^n + U_{i-1}^n\right). \tag{13.10}$$

A so-called, von Neumann stability analysis, shows that the scheme is stable as long as $\Delta t \le \frac{\Delta x^2}{2\alpha}$.

Solution. Program_13c.py uses equation (13.10) to give Figure 13.3(b). The reader is encouraged to change the parameters Δx and α and see how it affects stability. The program lists both the non-vectorized (using loops) and the vectorized commands.

Partial Derivative	FDA	Order and Type
$\dfrac{\partial U}{\partial x} = U_x$	$\dfrac{U_{i+1}^n - U_i^n}{\Delta x}$	First Order Forward
$\dfrac{\partial U}{\partial x} = U_x$	$\dfrac{U_i^n - U_{i-1}^n}{\Delta x}$	First Order Backward
$\dfrac{\partial U}{\partial x} = U_x$	$\dfrac{U_{i+1}^n - U_{i-1}^n}{2\Delta x}$	Second Order Central
$\dfrac{\partial^2 U}{\partial x^2} = U_{xx}$	$\dfrac{U_{i+1}^n - 2U_i^n - U_{i-1}^n}{\Delta x^2}$	Second Order Symmetric
$\dfrac{\partial U}{\partial t} = U_t$	$\dfrac{U_i^{n+1} - U_i^n}{\Delta t}$	First Order Forward
$\dfrac{\partial U}{\partial t} = U_t$	$\dfrac{U_i^n - U_i^{n-1}}{\Delta t}$	First Order Backward
$\dfrac{\partial U}{\partial t} = U_t$	$\dfrac{U_i^{n+1} - U_i^{n-1}}{2\Delta t}$	Second Order Central
$\dfrac{\partial^2 U}{\partial t^2} = U_{tt}$	$\dfrac{U_i^{n+1} - 2U_i^n - U_i^{n-1}}{\Delta t^2}$	Second Order Symmetric

Table 13.1 Finite difference approximations (FDAs) for partial derivatives derived from Taylor series expansions.

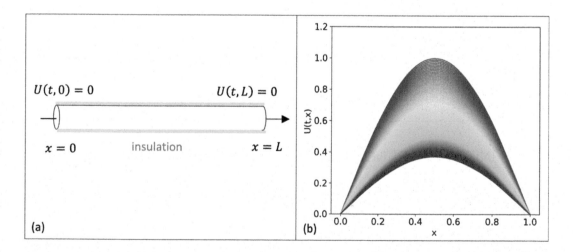

Figure 13.3 (a) Heat diffusion in an insulated rod, $U(t,0) = U(t,L) = 0$, $U(0,x) = \sin\left(\frac{\pi x}{L}\right)$, $\alpha = 0.1$, $L = 1$, for equation (13.9). (b) Numerical solution $U(t,x)$, for $0 \le t \le 1$, is stable as $\Delta t = 0.001$, and $\frac{\Delta x^2}{2\alpha} = 0.002$.

```
# Program_13c.py: The Heat Equation. Modeling heat in a rod.
import numpy as np
import matplotlib.pyplot as plt
import matplotlib.cm as cm
nt , nx , tmax , xmax , alpha = 1001 , 51 , 1 , 1 , 0.1
dt , dx = tmax/(nt-1) , xmax/(nx-1)
U , x = np.zeros((nx,nt)) , np.linspace(0, xmax, nx) # Initialize.
U[:,0] = np.sin(np.pi * x)                 # Initial conditions.
U[0,:] = U[-1,:] = 0                        # Boundary conditions.
r = alpha * (dt / dx ** 2)
# Vectorized solution.
for n in range(nt-1):
    U[1:-1,n+1]=U[1:-1,n]+r*(U[2:,n]-2*U[1:-1,n]+U[:-2,n])
# Non-vectorized commands using loops.
# for n in range(0,nt-1):
#   for i in range(0,nx-1):
#     U[0,n+1]=0
#     U[i,n+1] = U[i,n] + alpha*(dt/dx**2.0)* \
#     (U[i+1,n]-2.0*U[i,n]+U[i-1,n])
# for i in range(0 , nx):
#   x[i] = i * dx
plt.figure()
plt.rcParams["font.size"] = "16"
initial_cmap = cm.get_cmap("rainbow")
reversed_cmap=initial_cmap.reversed()
color=iter(reversed_cmap(np.linspace(0 , 10 , nt)))
for i in range(0 , nt , 10):
  c=next(color)
  plt.plot(x , U[:,i] , c = c)
plt.xlabel("x")
plt.ylabel("U(t,x)")
plt.ylim([0 , 1.2])
plt.show()
```

13.4 FINITE DIFFERENCE METHOD: THE WAVE EQUATION

This final section is concerned with a hyperbolic PDE used to model the oscillation of a guitar string.

Example 13.4.1. Let $U(t,x)$ represent the displacement of the string at distance x and time t. The one-dimensional wave equation used to model the vibration of a string is given by

$$U_{tt} - c^2 U_{xx} = 0, \tag{13.11}$$

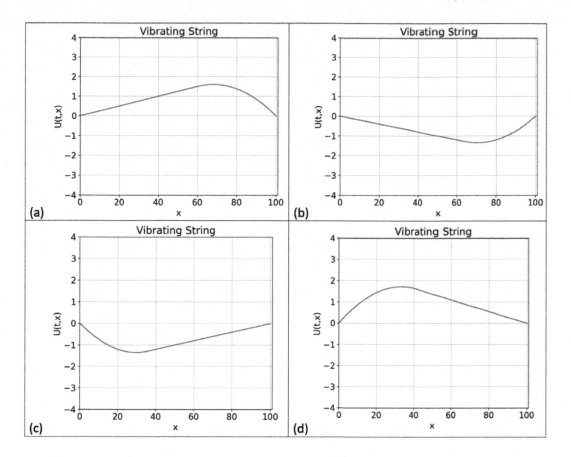

Figure 13.4 Vibration of a guitar string. Screenshots from the animation produced by running Program13d.py. The string vibrates in an oscillatory manner, (a)–(d), and back to (a).

where $c^2 = \frac{T_0}{\rho}$, T_0 is tension, and ρ is density. Using the symmetric differences in both time and space FDAs, show that

$$U_{i+1}^n = 2U_i^n - U_i^{n-1} + \frac{c^2 \Delta t^2}{\Delta x^2} \left(U_{i+1}^n - 2U_i^n + U_{i-1}^n \right). \qquad (13.12)$$

Solution. Program_13d.py employs equation (13.12) to give an animation with screenshots shown in Figure 12.4. The length of the string is $L = 100$, and the initial displacement is given by $U[x, 0] = U[i, 0]$ in the program. In this problem, take $c = 1$, $U_t(x, 0) = 0$, $U(0, t) = 0$ and $U(L, t) = 0$.

```
# Program_13d.py: The wave equation. Vibrating string.
import numpy as np
from matplotlib import pyplot as plt
from matplotlib import animation
U = np.zeros((101 , 3) , float)
k = range(0 , 101)
```

```
c=1
def init():                       # U[x,0] initial condition.
    for i in range(0, 81):
        U[i, 0] = 0.00125*i
        for i in range (81, 101):
            U[i, 0] = 0.1 - 0.005*(i - 80)
def animate(num):
    for i in range(1, 100):
        U[i,2]=2*U[i,1]-U[i,0]+c**2*(U[i+1,1]-2*U[i,1]+U[i-1,1])
    line.set_data(k,U[k,2])
    for m in range (0,101):
        U[m, 0] = U[m, 1]
        U[m, 1] = U[m, 2]
    return line
fig = plt.figure()
ax=fig.add_subplot(111,autoscale_on=False,xlim=(0,101),ylim=(-4,4))
ax.grid()
plt.title("Vibrating String")
line, = ax.plot(k, U[k,0], lw=2)
plt.xlabel("x")
plt.ylabel("U(t,x)")
for i in range(3,100):
    U[i,2] = 2*U[i,1]-U[i,0]+c**2*(U[i+1,0]-2*U[i,0]+U[i-1,0] )
anim = animation.FuncAnimation(fig, animate,init_func=init,frames= \
            np.linspace(0,10,1000),interval=10,blit=False)
plt.show()
```

EXERCISES

13.1 The Euler method may be modified to eliminate stability problems. Carry out a literature search on the backward Euler method. This method is an implicit method, the iterative formula is given by:

$$y_{n+1} = y_n + hf\left(x_{n+1}, y_{n+1}\right),$$

and it is implicit as there are y_{n+1} terms on both sides of the iterative formula. Use the implicit Euler method to solve the IVP:

$$\frac{dy}{dx} = (x - 3.2)y + 8x \exp\left(\frac{(x - 3.2)^2}{2}\right) \cos\left(4x^2\right), \qquad (P13.1)$$

given that $x_0 = 0, y_0 = 1 h = 0.1$ and $0 \le x \le 5$. When using the implicit method, the Newton-Raphson method (see Chapter 5) can be used to solve the implicit equations. Plot the solutions and comment on the stability as h increases from $h = 0.1$ to $h = 0.5$. To check your answers, the analytical solution is shown in Figure 13.5.

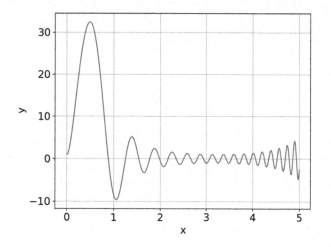

Figure 13.5 Analytical solution to the IVP (P13.1) with $x_0 = 0, y_0 = 1, h = 0.1$ and $0 \leq x \leq 5$.

13.2 Research the implicit Adams-Bashforth method using books and the internet. The two-step Adams-Bashforth method uses the iterative formula:

$$x_{n+2} = x_{n+1} + \frac{h}{2} \left(3f\left(t_{n+1}, x_{n+1}\right) - f\left(t_n, x_n\right) \right).$$

Use this iterative formula, and the RK4 method to compute initial steps, to plot the solution to the IVP (13.10), with $x_0 = 0, y_0 = 1, h = 0.1$ and $0 \leq x \leq 5$.

13.3 The simplest form of the linear advection equation (hyperbolic PDE) is

$$U_t + vU_x = 0, \tag{P13.2}$$

where v is the advection velocity. This equation models pure advection along a one-dimensional channel and it is simple to calculate an exact solution. Numerical schemes can be checked for stability when modeling more complicated hyperbolic PDEs. The first-order upwind scheme can be obtained using FDA (13.9) for both time and displacement. Show that the advection equation can be written as

$$U_i^{n+1} = U_i^n - v\frac{\Delta t}{\Delta x}\left(U_i^n - U_{i-1}^n\right).$$

Given that $U(x,0) = \exp\left(-200(x - 0.25)^2\right)$, and the exact solution is $\exp\left(-200(x - vt - 0.25)^2\right)$, use Python to produce exact and numerical solutions when the length of the channel is $x = 1$, and $v = 0.5$. Use 101 nodes, so $\Delta x = 0.01$, and $\Delta t = 0.01$. Take ghost nodes $U_{-1} = U_1$ and $U_N = U_{N-2}$, so the numerical scheme works. Investigate ways to improve the results. Numerical solutions for $t = 0.25, 0.5, 0.75$ and $t = 1$ seconds, are shown in Figure 13.6.

13.4 The two-dimensional heat diffusion equation is given by

$$U_t = \alpha \left(U_{xx} + U_{yy} \right),$$

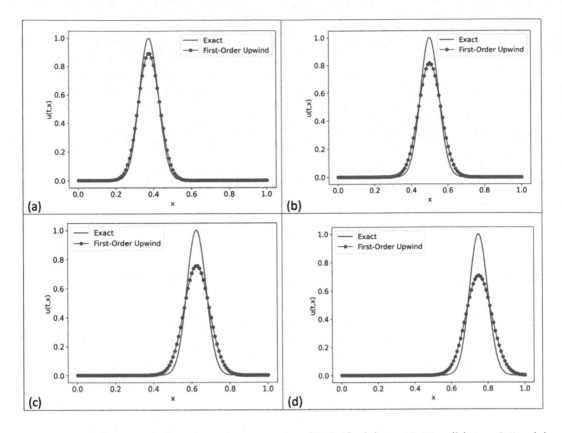

Figure 13.6 Solution of the advection equation (P13.2). (a) $t = 0.25s$; (b) $t = 0.5s$; (c) $t = 0.75s$; (d) $t = 1s$.

using a forward difference in time and symmetric differences in both spatial directions, show that

$$U_{i,j}^{n+1} = U_{i,j}^n + \frac{\alpha \Delta t}{\Delta x^2} \left(U_{i-1,j}^n - 2U_{i,j}^n + U_{i+1,j}^n \right) + \frac{\alpha \Delta t}{\Delta y^2} \left(U_{i,j-1}^n - 2U_{i,j}^n + U_{i,j+1}^n \right).$$

Solve the problem numerically given, the heat diffusion parameter, $\alpha = 0.01$, $\Delta x = \Delta y = 0.02$, take 50 nodes in both the x and y directions, and use the boundary conditions shown in Figure 13.7(a). Write a program using for loops for the iterations, and also write the program in vectorized form. Compare the running time for both programs. The solution is shown in Figure 13.7(b) for $0 \le t \le 5$.

Solutions to the Exercises may be found here:

https://drstephenlynch.github.io/webpages/Solutions_Section_2.html.

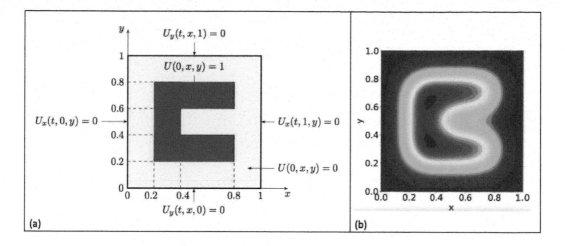

Figure 13.7 (a) Boundary conditions for heat diffusion across a metal sheet. (b) Numerical solution using Python.

FURTHER READING

[1] DuChateau, P. and Zachmann, D. (2011). *Schaum's Outline of Partial Differential Equations*. McGraw Hill, New York.

[2] Gottlieb, S. (2011). *Strong Stability Preserving Runge-Kutta and Multistep Time Discretizations*. World Scientific, Singapore.

[3] Kong, Q. Siauw, T. and Bayen, A. (2020). *Python Programming and Numerical Methods. A Guide for Engineers and Scientists*. Academic Press, Cambridge, MA.

[4] Lambers, J.V., Sumner-Mooney, A.C. and Montiforte, V.A. (2020). *Explorations in Numerical Analysis: Python Edition*. World Scientific, Singapore.

[5] Linge, S. and Langtangen, H.P. (2018). *Programming for Computations - Python: A Gentle Introduction to Numerical Simulations with Python*. Springer, New York.

[6] Olver, P.J. (2016). *Introduction to Partial Differential Equations*. Springer, New York.

## CHAPTER 14

Physics

Physics is the branch of science investigating the nature and properties of matter and energy. It includes subjects such as astrophysics, electricity and magnetism, general and special theories of relativity, heat, mechanics, optics, signal processing and quantum mechanics, for example. It is the closest subject aligned to mathematics. The first section introduces the fast Fourier transform (fft). In this book, it is used in signal processing and as a chaos detector, but it has far more applications in the real world, and is considered by many to be the most powerful tool in the whole of mathematics. The second section, from nonlinear optics, presents a simple example of how a complex iterative map may be used to model the propagation of light waves through nonlinear optical fiber. Next, equations that model the Josephson junction, a superconducting nonlinear threshold oscillator, are described. Finally, the motion of three celestial bodies in three-dimensional space is described using a system of ODEs, which can be solved numerically using Euler's method.

Aims and Objectives:

- To introduce the fast Fourier transform (fft).

- To model the propagation of light through nonlinear optical fiber.

- To investigate a Josephson junction.

- To model the motion of celestial bodies.

At the end of the chapter, the reader should be able to

- use the fft in signal processing and as a chaos detector;

- use a complex iterative map to model the propagation of light in a simple fiber ring resonator;

- explain the dynamics of a Josephson junction;

- model the motion of celestial bodies.

DOI: 10.1201/9781003285816-14

14.1 THE FAST FOURIER TRANSFORM

The Fourier transform is a mathematical transform with many applications in science, and in particular, physics.

Definition 14.1.1. The continuous Fourier transform is defined by

$$F(\omega) = \int_{-\infty}^{\infty} f(t)e^{-2\pi i\omega t}dt,$$

which transforms a mathematical function of time, $f(t)$, into a function of frequency, $F(\omega)$. The new function is the Fourier transform or the Fourier spectrum of the function f.

Definition 14.1.2. The inverse Fourier transform is defined by

$$f(t) = \int_{-\infty}^{\infty} F(\omega)e^{2\pi i\omega t}d\omega.$$

The continuous Fourier transform converts an infinitely long time-domain signal into a continuous spectrum of an infinite number of sinusoidal curves. In many physical applications, scientists deal with discretely sampled signals, usually at constant intervals. For such data, the discrete Fourier transform is appropriate.

Definition 14.1.3. The discrete Fourier transform and its inverse for vectors of length N are defined by

$$X_k = \sum_{n=1}^{N} t_n \omega_N^{(n-1)(k-1)},$$

and

$$x_n = \frac{1}{N}\sum_{k=1}^{N} X_k \omega_N^{-(n-1)(k-1)},$$

where

$$\omega_N = e^{(-2\pi i)/N},$$

and each X_k is a complex number that encodes both amplitude and phase of a sinusoidal component of function x_n.

A fast Fourier transform, or fft, is an algorithm to compute the discrete Fourier transform. The fft was first discovered by Gauss in 1805 but the modern incarnation is attributed to Cooley and Tukey [4] in 1965. Computing a set of N data points using the discrete Fourier transform requires $\mathcal{O}\left(N^2\right)$ arithmetic operations, whilst a fft can compute the same discrete Fourier transform in only $\mathcal{O}(N \log N)$ operations. There is an fft function in Python for finding the fft.

The fft is a powerful signal analysis tool, applicable to a wide variety of fields including acoustics, applied mechanics, communications, digital filtering, instrumentation, medical imaging, modal analysis, numerical analysis, seismography and spectral analysis, for example.

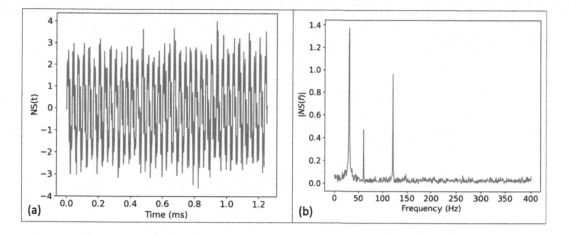

Figure 14.1 (a) Noisy signal, $NS(t) = noise + 0.5\sin(2\pi(60t)) + 1\sin(2\pi(120t)) + 2\sin(2\pi(30t))$. (b) The amplitude spectrum $|NS(f)|$. You can read off the amplitudes and frequencies. Increase the number of sampling points, Ns, in the program to increase accuracy.

Example 14.1.1. A common use of Fourier transforms is to find the frequency components of a signal buried in a noisy time domain signal. Consider data sampled at $800Hz$. Take a signal containing a $60Hz$ sinusoid of amplitude 0.5, a $120Hz$ sinusoid of amplitude 1, and a $30Hz$ sinusoid of amplitude 2. Corrupt it with some zero-mean random noise. Use Python to plot a graph of the signal and write a program that plots an amplitude spectrum for the signal.

Solution. Figure 14.1(a) shows the sum of three sinusoids corrupted with zero-mean random noise, $NS(t)$ say, and Figure 14.1(b) displays the amplitude spectrum of $NS(t)$. The program for plotting the figures is listed in Program_14a.py.

```
# Program_14a.py: Fast Fourier transform of a noisy signal.
import numpy as np
import matplotlib.pyplot as plt
from scipy.fftpack import fft
Ns = 1000                       # Number of sampling points
Fs = 800                        # Sampling frequency
T = 1/Fs                        # Sample time
t = np.linspace(0, Ns*T, Ns)
amp1, amp2 , amp3 = 0.5 , 1 , 2
freq1, freq2 , freq3 = 60 , 120 , 30
# Sum a 30Hz, 60Hz and 120 Hz sinusoid
x = amp1 * np.sin(2*np.pi * freq1*t) + amp2*np.sin(2*np.pi * freq2*t) \
    +amp3 * np.sin(2*np.pi * freq3*t)
NS = x + 0.5*np.random.randn(Ns)          # Add noise.
fig1 = plt.figure()
plt.plot(t, NS)
plt.xlabel("Time (ms)", fontsize=15)
plt.ylabel("NS(t)", fontsize=15)
```

```
plt.tick_params(labelsize=15)
fig2 = plt.figure()
Sf = fft(NS)
xf = np.linspace(0, 1/(2*T), Ns//2)
plt.plot(xf, 2/Ns * np.abs(Sf[0:Ns//2]))
plt.xlabel("Frequency (Hz)", fontsize=15)
plt.ylabel("$|NS(f)|$", fontsize=15)
plt.tick_params(labelsize=15)
plt.show()
```

Readers interested in signal processing with Python are directed to references [3], [7] and [13]. The fft can also be used as a chaos detector, an example in given in the exercises at the end of the chapter.

14.2   THE SIMPLE FIBER RING (SFR) RESONATOR

Nonlinear optics is a subject within physics and is the study of light in nonlinear media. For an in-depth introduction, readers are directed to [2], and for applications, Agrawal's book [1] provides plenty of examples. One major application of nonlinear optics is concerned with bistable devices, which are devices displaying hysteresis. For an introduction to nonlinear optics using Python, readers can see [8]. This section is concerned with the nonlinear simple fiber ring (SFR) resonator, as depicted in Figure 14.2(a), which is a well-known bistable device. The simple complex iterative map used to model the system was derived by Ikeda et al. [9] when they simplified a system of delayed differential equations into a complex iterative map. Those interested in the use of delay differential equations in nonlinear optics can read our paper on delayed-dynamical bistability within and without the rotating wave approximation [11]. The complex iterative equation derived by Ikeda is known as the Ikeda map

$$E_{n+1} = A + BE_n e^{i|E_n|^2}, \tag{14.1}$$

where E_n is the electric field strength of the light on the n'th loop, $A = E_{in}$ is related to input power, and $0 \le B \le 1$, gives a measure of the fiber coupling ratio or, how much light circulates in the loop. In the program, $|A|^2$ gives the input power, and $|E_n|^2$ is the output power.

Example 14.2.1. Use Python to plot a bifurcation diagram for the Ikeda map (14.1) as the input power $|A|^2$, is ramped up to $|A|^2 = 10$, and then ramped down again to zero, when $B = 0.15$, and 15% of the light circulates in the loop.

Solution. Figure 14.2(b) shows the bifurcation diagram. There is steady-state (period-one) behavior, hysteresis, and period-doubling and period-undoubling into and out of chaos. As the bistable (hysteresis) loop is isolated from instabilities, this device could be used in applications. In the program, $1j = \sqrt{-1}$.

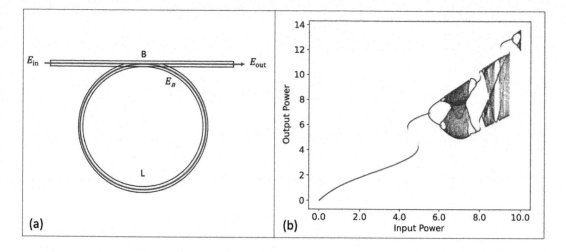

Figure 14.2 (a) The nonlinear simple fiber ring (SFR) resonator constructed with optical fiber. The red curves depict the laser in the optical fiber. (b) Bifurcation diagram of the Ikeda map (14.1) with feedback. There is a counterclockwise hysteresis loop and period-doubling and period-undoubling in and out of chaos. The red points indicate ramp up power, and the blue points represent ramp down power.

```
# Program_14b.py: Bifurcation diagram of the Ikeda map.
from matplotlib import pyplot as plt
import numpy as np
B , phi , Pmax , En = 0.15 , 0 , 10 , 0    # phi is a linear phase shift.
half_N = 99999
N = 2*half_N + 1
N1 = 1 + half_N
esqr_up, esqr_down = [], []
ns_up = np.arange(half_N)
ns_down = np.arange(N1, N)
# Ramp the power up
for n in ns_up:
    En = np.sqrt(n * Pmax / N1) + B * En * np.exp(1j*((abs(En))**2 - phi))
    esqr1 = abs(En)**2
    esqr_up.append([n, esqr1])
esqr_up = np.array(esqr_up)
# Ramp the power down
for n in ns_down:
    En = np.sqrt(2 * Pmax - n * Pmax / N1) + \
    B*En* np.exp(1j*((abs(En))**2 - phi))
    esqr1 = abs(En)**2
    esqr_down.append([N-n, esqr1])
esqr_down=np.array(esqr_down)
fig, ax = plt.subplots()
xtick_labels = np.linspace(0, Pmax, 6)
ax.set_xticks([x / Pmax * N1 for x in xtick_labels])
ax.set_xticklabels(["{:.1f}".format(xtick) for xtick in xtick_labels])
plt.plot(esqr_up[:, 0], esqr_up[:, 1], "r.", markersize=0.1)
```

```
plt.plot(esqr_down[:, 0], esqr_down[:, 1], "b.", markersize=0.1)
plt.xlabel("Input Power", fontsize=15)
plt.ylabel("Output Power", fontsize=15)
plt.tick_params(labelsize=15)
plt.show()
```

14.3   THE JOSEPHSON JUNCTION

In 1962, Brian Josephson predicted that, for very low temperatures (typically 4 Kelvin), pairs of superconducting electrons could tunnel through an insulating layer from one superconductor to another with no energy loss [10]. For a history of this device, the theory, and a list of its many applications, readers are directed to [15]. They have important applications in quantum-mechanical circuits, superconducting quantum interference devices (SQUIDS), which act as very sensitive magnetometers, and rapid single flux quantum (RSFQ) digital circuits. For future reference in relation to Section III of the book, Josephson junctions (JJs) are also natural threshold oscillators and can act like biological neurons, see [6] and [12], for example. The dimensionless differential equation used to model a resistively shunted JJ [6] is written as:

$$\frac{d^2\phi}{d\tau^2} + \beta_J \frac{d\phi}{d\tau} + \sin\phi = \kappa, \tag{14.2}$$

where ϕ is a phase difference, β_J is a parameter inversely related to the Josephson plasma frequency, ω_J, κ represents the total current across the junction, and $\frac{d\phi}{dt} = \omega_J \frac{d\phi}{d\tau}$. Suppose that, $\frac{d\phi}{d\tau} = \Omega$, then the second order ODE (14.2) can be written as a system of two first-order ODEs:

$$\frac{d\phi}{d\tau} = \Omega, \quad \frac{d\Omega}{d\tau} = \kappa - \beta_J\Omega - \sin\phi, \tag{14.3}$$

where $\eta = \beta_J\Omega$, is proportional to voltage.

Example 14.3.1. Show that the resistively shunted JJ acts as a threshold oscillator given that, $\beta_J = 1.2$, and κ increases, from $\kappa = 0.1$ to $\kappa = 2$. Use Python to produce an animation of the bifurcation of the limit cycle.

Solution. Program_14c.py produces an animation of the phase portrait of the resistively shunted JJ, see equations (14.3), as current across the junction increases. Figure 14.3 shows snapshots of the animation.

```
# Program_14c.py: Animation of a JJ limit cycle bifurcation.
from matplotlib import pyplot as plt
from matplotlib import animation
import numpy as np
from scipy.integrate import odeint
from matplotlib import style
fig=plt.figure()
```

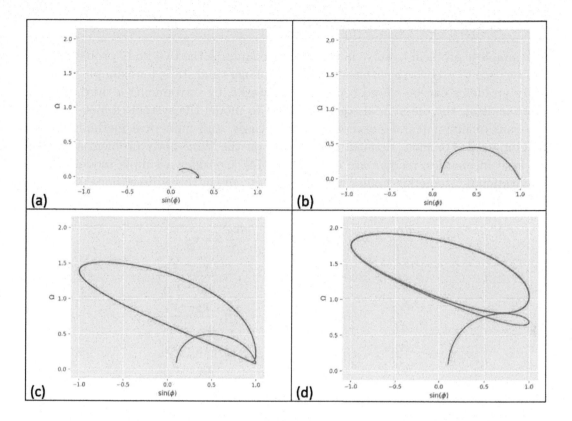

Figure 14.3 Animation of a resistively shunted JJ acting as a threshold oscillator. The current across the junction increases from $\kappa = 0.1$ to $\kappa = 2$. (a) No oscillation, for small κ. (b) Close to threshold. (c) Bifurcation of a limit cycle at $\kappa \approx 1.0025$. (d) The limit cycle moves vertically upwards for $\kappa > 1.0025$.

```
myimages=[]
BJ=1.2;Tmax=100;
def JJ_ODE(x, t):
    return [x[1],kappa-BJ*x[1]-np.sin(x[0])]
style.use("ggplot")          # To give oscilloscope-like graph.
time = np.arange(0, Tmax, 0.1)
x0=[0.1,0.1]
for kappa in np.arange(0.1, 2, 0.1):
    xs = odeint(JJ_ODE, x0, time)
    imgplot = plt.plot(np.sin(xs[:,0]), xs[:,1], "g-")
    myimages.append(imgplot)
my_anim=animation.ArtistAnimation(fig,myimages,interval=500,\
                          blit=False,repeat_delay=100)
plt.rcParams["font.size"] = "18"
plt.xlabel("$\sin(\phi)$")
plt.ylabel("$\Omega$")
plt.show()
```

14.4 MOTION OF PLANETARY BODIES

The simplest problem to solve in classical mechanics is the two-body problem, which can be used to predict the orbits of satellites, planets and stars, for example. The two-body problems can be solved completely, however, by introducing a third body (or more bodies), the differential equations used to model the system cannot be solved in terms of first integrals, except in special cases, and numerical methods have to be employed. For more detailed information on the three-body problem applied to celestial mechanics, readers are directed to [14]. Consider a simple model of three gravitational interacting bodies with masses, m_i say, in three-dimensional space (see Figure 14.4(a)), which can be modeled using Newton's equations of motion:

$$\frac{d\mathbf{r_1}}{dt} = \mathbf{v_1}, \quad \frac{d\mathbf{r_2}}{dt} = \mathbf{v_2}, \quad \frac{d\mathbf{r_3}}{dt} = \mathbf{v_3},$$

$$\frac{d\mathbf{v_1}}{dt} = -Gm_2\frac{\mathbf{r_1} - \mathbf{r_2}}{|\mathbf{r_1} - \mathbf{r_2}|^3} - Gm_3\frac{\mathbf{r_1} - \mathbf{r_3}}{|\mathbf{r_1} - \mathbf{r_3}|^3},$$

$$\frac{d\mathbf{v_2}}{dt} = -Gm_3\frac{\mathbf{r_2} - \mathbf{r_3}}{|\mathbf{r_2} - \mathbf{r_3}|^3} - Gm_1\frac{\mathbf{r_2} - \mathbf{r_1}}{|\mathbf{r_2} - \mathbf{r_1}|^3},$$

$$\frac{d\mathbf{v_3}}{dt} = -Gm_1\frac{\mathbf{r_3} - \mathbf{r_1}}{|\mathbf{r_3} - \mathbf{r_1}|^3} - Gm_2\frac{\mathbf{r_3} - \mathbf{r_2}}{|\mathbf{r_3} - \mathbf{r_2}|^3}, \quad (14.4)$$

where G is a gravitational constant, $\mathbf{r_i}$ and $\mathbf{v_i}$ are the position and velocity vectors of masses m_i, respectively.

Example 14.4.1. Use Euler's numerical method to solve the ODEs (14.4) given that, $G = 1, m_1 = 100, m_2 = 1, m_3 = 1$, $\mathbf{r_1}(0) = (0,0,0)$, $\mathbf{v_1}(0) = (0,0,0)$ and

(i) $\mathbf{r_2}(0) = (1,0,0)$, $\mathbf{v_2}(0) = (0,10,0)$; $\mathbf{r_3}(0) = (-1,0,0)$, $\mathbf{v_3}(0) = (0,-10,0)$;

(ii) $\mathbf{r_2}(0) = (1,0,1)$, $\mathbf{v_2}(0) = (0,10,0)$; $\mathbf{r_3}(0) = (-1,0,1)$, $\mathbf{v_3}(0) = (0,-10,0)$;

(iii) $\mathbf{r_2}(0) = (1,0.1,0)$, $\mathbf{v_2}(0) = (1,9,-1)$; $\mathbf{r_3}(0) = (-1,0,0)$, $\mathbf{v_3}(0) = (0,-10,0)$.

Solution. Program_14d.py models three bodies using equations (14.4). The trajectories of the orbits are displayed in Figure 14.4. Note that in Spyder, you can rotate the three-dimensional images by simply clicking on the image and moving the mouse.

```
# Program_14d.py: Three Body Problem. Case (iii).
import numpy as np
import matplotlib.pyplot as plt
from numpy.linalg import norm
fig = plt.figure(figsize=(10, 10))
ax = fig.add_subplot(111, projection="3d")
xmax = 2
def Planets(r1,r2,r3,v1,v2,v3, G=1,m1=100,m2=1,m3=1):
```

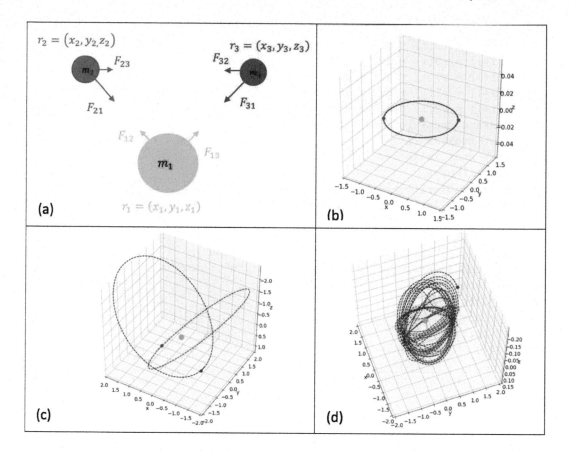

Figure 14.4 The three-body problem. (a) The positions and forces on the three bodies. (b) Circular motion for equations (14.4) under conditions (i). The red and blue bodies move on circular orbits about the heavier orange body in a plane. (c) Elliptic motions for equations (14.4) under conditions (ii). (d) What appears to be more random behavior under conditions (iii).

```
    r12 , r13 , r23 = norm(r1-r2) , norm(r1-r3) , norm(r2-r3)
    dr1 , dr2 , dr3  = v1 , v2 , v3
    dv1 = -G*m2*(r1-r2)/r12**3-G*m3*(r1-r3)/r13**3
    dv2 = -G*m3*(r2-r3)/r23**3-G*m1*(r2-r1)/r12**3
    dv3 = -G*m1*(r3-r1)/r13**3-G*m2*(r3-r2)/r23**3
    return dr1 , dr2 , dr3 , dv1 , dv2 , dv3
n , dt = 800000 , 0.00001
r1,r2,r3=np.zeros((n,3)),np.zeros((n,3)),np.zeros((n,3))
v1,v2,v3=np.zeros((n,3)),np.zeros((n,3)),np.zeros((n,3))
# The initial conditions.
r1[0],r2[0],r3[0]=np.array([0,0,0]),np.array([1,0.1,0]),np.array([-1,0,0])
v1[0],v2[0],v3[0]=np.array([0,0,0]),np.array([1,9,-1]),np.array([0,-10,0])
# Solve using Euler's numerical method.
# Use other numerical methods to increase accuracy.
for i in range(n-1):
    dr1,dr2,dr3,dv1,dv2,dv3 = Planets(r1[i],r2[i],r3[i],v1[i],v2[i],v3[i])
```

```
    r1[i+1] = r1[i] + dr1 * dt
    r2[i+1] = r2[i] + dr2 * dt
    r3[i+1] = r3[i] + dr3 * dt
    v1[i+1] = v1[i] + dv1 * dt
    v2[i+1] = v2[i] + dv2 * dt
    v3[i+1] = v3[i] + dv3 * dt
plt.rcParams.update({"font.size": 14})
plt.xlim(-xmax,xmax)
plt.ylim(-xmax,xmax)
fig = plt.figure()
ax.plot(r1[:,0],r1[:,1],r1[:,2],"--",color="orange")
ax.scatter3D(r1[n-1,0],r1[n-1,1],r1[n-1,2],"o",color="orange",s=200)
ax.plot(r2[:,0],r2[:,1],r2[:,2],"--",color="red")
ax.scatter3D(r2[n-1,0],r2[n-1,1],r2[n-1,2],"o",color="red",s=50)
ax.plot(r3[:,0],r3[:,1],r3[:,2],"--",color="blue")
ax.scatter3D(r3[n-1,0],r3[n-1,1],r3[n-1,2],"o",color="blue",s=50)
ax.set_xlabel("x")
ax.set_ylabel("y")
ax.set_zlabel("z")
ax.view_init(-120, 20)
plt.show()
```

EXERCISES

14.1 The power spectrum is given as $|\text{fft}(X)|^2$, where X is a vector of length n. Consider the 2-dimensional discrete map defined by

$$x_{n+1} = 1 + \beta x_n - \alpha y_n^2$$
$$y_{n+1} = x_n, \tag{P14.1}$$

where α and β are constants. Suppose that $\alpha = 1$, plot iterative plots, and plots of log(power) against frequency for system (P14.1) when (i) $\beta = 0.05$ (periodic); (ii) $\beta = 0.12$ (quasi-periodic) and (iii) $\beta = 0.3$ (chaotic). In this case, the power spectra gives an indication as to whether or not the system is behaving chaotically.

14.2 Convert the one-dimensional complex Ikeda map (14.1), into a real two-dimensional mapping by writing $E_n = x_n + iy_n$. You are splitting E_n into its real and imaginary parts. Use Python to produce an animation of the (x_n, y_n) phase portrait as the parameter $|A|^2$ increases from $|A|^2 = 0$ to $|A|^2 = 10$. Describe how this relates to the bifurcation diagram, Figure 14.2(b).

14.3 The tunnelling JJ, see equations (14.3), also displays hysteresis, as shown in Figure 14.5, which shows a typical current-voltage, κ-$\langle \eta \rangle$, characteristic curve for a resistively shunted JJ, as the average voltage, $\langle \eta \rangle$, is increased and decreased. Use Python to produce Figure 14.5.

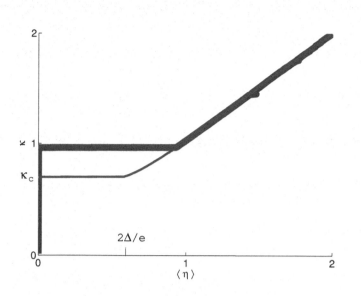

Figure 14.5 A typical current-voltage characteristic curve for the resistively shunted JJ. There is a clockwise hysteresis cycle. Note that the average voltage, $\langle \eta \rangle = \beta_J \langle \Omega \rangle$.

14.4 Consider the special case of two heavy bodies with masses m_1, m_2, and a third lighter body of mass m_3, all restricted to move in the plane. Bodies with masses m_1 and m_2, interact via gravity and rotate around each other, unperturbed by the body of mass m_3, which feels the full gravitational force of the heavier bodies. The ODEs to model the system are given by:

$$\frac{d^2\mathbf{r_1}}{dt^2} = \frac{-m_2^3}{(m_1 + m_2)^2} \frac{\mathbf{r_1}}{|\mathbf{r_1}|^3},$$

$$\mathbf{r_2} = -\frac{m_1}{m_2}\mathbf{r_1},$$

$$\frac{d^2\mathbf{r_3}}{dt^2} = -m_1 \frac{\mathbf{r_{31}}}{|\mathbf{r_{31}}|^3} - m_2 \frac{\mathbf{r_{32}}}{|\mathbf{r_{32}}|^3}, \tag{P14.2}$$

where $\mathbf{r_{ij}} = \mathbf{r_i} - \mathbf{r_j}$, and $\frac{d\mathbf{r_i}}{dt} = \mathbf{v_i}$. Use Python to plot the trajectories of the three bodies given that, $m_1 = 1$, $m_2 = 2$, $\mathbf{r_1}(0) = (1,0)$, $\mathbf{v_1}(0) = (0,1)$, $\mathbf{r_3}(0) = (1.5, 0)$, $\mathbf{v_3}(0) = (0, 2.5)$ and $t = 30$. This could be modeling an earth-moon-comet, system.

Solutions to the Exercises may be found here:

https://drstephenlynch.github.io/webpages/Solutions_Section_2.html.

FURTHER READING

[1] Agrawal, G. (2020). *Applications of Nonlinear Fiber Optics, 3rd Ed.* Academic Press, Cambridge, MA.

[2] Boyd, R.W. (2020). *Nonlinear Optics, 4th Ed.* Academic Press, Cambridge, MA.

[3] Charbit, M. (2017). *Digital Signal Processing (DSP) with Python Programming*, Wiley, New York.

[4] Cooley, J.W. and Tukey J.W. (1965). An algorithm for the machine calculation of complex Fourier series, *Math. Comput.* **19**, 297–301.

[5] Crotty, P., Schult, D. and Segall, K. (2010). Josephson junction simulation of neurons. *Phys. Rev.* **82**, 011914.

[6] Mishra, A., Ghosh, S., Dana, S.K. et al. (2021). Neuron-like spiking and bursting in Josephson junctions: A Review. *Chaos: An Interdiciplinary Journal of Nonlinear Science.* **31**, 052101.

[7] Haslwanter, T. (2021). *Hands-on Signal Analysis with Python: An Introduction*, Springer, New York.

[8] Kuzyk, M.G. (2017). *Nonlinear Optics: A Student's Perspective: With Python Problems and Examples.* CreateSpace Independent Publishing Platform, Scotts Valley, CA.

[9] Ikeda, K., Dido, H. and Akimoto, O. (1980). Optical turbulence: chaotic behavior of transmitted light. *Phys. Rev. Lett.* **45**(9), 709–712.

[10] Josephson, B.D. (1962). Possible new effects in superconductive tunnelling. *Physics Letters.* **1**(7), 251–253.

[11] Lynch, S., Alharbey, R.A., Hassan, S.S. and Batarfi, H.A. (2015). Delayed-dynamical bistability within and without the rotating wave approximation. *J. of Nonlinear Optical Physics and Materials* **24**(3), 1550037.

[12] Lynch, S., Borresen, J. and Latham, K. (2013). Josephson junction binary oscillator computing. *Proceedings of the IEEE International Superconductive Electronics Conference, Cambridge, MA.* DOI: 10.1109/ISEC.2013.6604275, 1-3.

[13] Unpingco, J. (2014). *Python for Signal Processing: Featuring IPython Notebooks.* Springer, New York.

[14] Valtonen, M., Anosova, J., Kholshevnikov, K. et al. (2016). *The Three-Body Problem from Pythagoras to Hawking.* Springer, New York.

[15] Wolf, E.L., Arnold, G.B., Gurvitch, M.A. and Zasadzinski, J.F. (2017). *Josephson Junctions: History, Devices, and Applications* Jenny Stanford Publishing, Singapore.

## CHAPTER 15

Statistics

Statistics is a branch of mathematics dealing with the collection, tabulation and interpretation of numerical data. There are specialist statistical software packages available such as GraphPad Prism, Minitab, Statistical Analysis Software (SAS), Statistical Package for the Social Sciences (SPSS) and R, for example. This chapter shows the reader how Python can be used to solve a wide range of statistical problems and is also useful in Data Science and Machine Learning. For more detailed information on how Python is used in statistics, see [3], [7] and [15]. The first section covers simple linear regression and shows the linear relationship between carbon dioxide emissions and gross domestic product in the USA from 2000 to 2020. The second section is concerned with stochastic processes and Markov chains, directed graphs are plotted and convergence to steady states are illustrated with plots. The student's t-test is discussed in the third section and a simple two sample t-test is applied to the heights of female and male students in a lecture. The final section introduces powerful Monte-Carlo simulations and a simple example is presented of gambling on a European roulette wheel, where it is clearly demonstrated that casinos and online betting companies always win in the long term, and why none of us should gamble.

Aims and Objectives:

- To show how Python can be used to solve problems in statistics.

At the end of the chapter, the reader should be able to

- Carry out linear regression on data sets.

- Draw Markov chains and plot stability diagrams.

- Carry out statistical hypotheses tests using the t-test.

- Apply Monte-Carlo simulations to a wide range of applications.

15.1 LINEAR REGRESSION

Linear regression is the most basic and commonly used tool in statistics, where a linear equation represents the relationship between two variables. There are two basic

DOI: 10.1201/9781003285816-15

types of linear regression, simple linear regression (SLR) and multiple linear regression (MLR). The simple MLR model takes the form:

$$\mathbf{y} = \mathbf{Xb} + \mathbf{e},$$

where

$$
\mathbf{y} = \begin{pmatrix} y_1 \\ y_2 \\ \vdots \\ y_n \end{pmatrix}, \quad
\mathbf{X} = \begin{pmatrix} 1 & x_{11} & \cdots & x_{1p} \\ 1 & x_{21} & \cdots & x_{2p} \\ \vdots & & & \\ 1 & x_{n1} & \cdots & x_{np} \end{pmatrix}, \quad
\mathbf{b} = \begin{pmatrix} b_0 \\ b_1 \\ \vdots \\ b_n \end{pmatrix}, \quad
\mathbf{e} = \begin{pmatrix} e_1 \\ e_2 \\ \vdots \\ e_n \end{pmatrix},
$$

where $\mathbf{Y}$ is the dependent variable, $\mathbf{x}$ is the independent variable, b_0 is the intercept, $\mathbf{b}$ is the slope, and $\mathbf{e}$ is the error term. This section will be concerned with a SLR model of the form:

$$y = f(x) = b_0 + b_1 x, \tag{15.1}$$

where the observed data satisfies the equations:

$$y_i = b_0 + b_1 x_i + e_i. \tag{15.2}$$

Figure 15.1(a) shows a typical computed regression line, the observed data points (blue +), the predicted responses (black +) and the residuals. The basic idea is to mimimize the sum of the squares of the residuals (SSR):

$$SSR = \sum_{i=1}^{n} e_i^2 = \sum_{i=1}^{n} (y_i - f(x_i))^2,$$

which is known as the method of ordinary least squares (OLS). The coefficient of determination, or R^2 value, is a goodness-of-fit statistic for regression analysis. Usually, larger values of R^2 indicate goodness-of-fit, however, there are many caveats, see [6] and [12], for example. There are two fantastic libraries for regression in Python, scikit-learn is widely used in machine learning and data science:

https://scikit-learn.org/stable/,

and covers regression, classification, clustering, dimensionality reduction, model selection and preprocessing, for example. For more in-depth analysis, the library statsmodels is a powerful package for estimation of statistical models, performing statistical tests, data exploration, and more:

https://www.statsmodels.org/stable/index.html.

Example 15.1.1. The file CO2_GDP_USA_Data.csv lists the data for the USA from 2000 to 2020, for carbon dioxide emissions per capita, column labelled "co2 per capita (metric tons)", and gross domestic product per capita, column labelled "gdp per

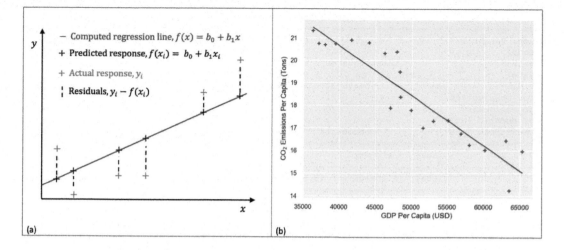

Figure 15.1 (a) Simple linear regression. (b) Simple linear regression for carbon dioxide emissions per capita, and gross domestic product per capita, for USA, in the years 2000 to 2020. In this case, the gradient $b_1 = -0.00022269$, the y-intercept is 29.567819, the mean squared error is 0.6 and the R^2 score is 0.86.

capita (USD)." Use Python to perform a linear regression analysis and plot the line of best fit for the data. Print out the results of OLS regression from the statsmodels library.

Solution. Program_15a.py uses the sklearn and the statsmodels libraries to produce Figures 15.1(b) and 15.2. The results are the same that would be produced using the statistics packages Minitab, R and SPSS, for example.

```
#Program_15a.py: Simple Linear Regression.
import matplotlib.pyplot as plt
import numpy as np
from sklearn import linear_model
from sklearn.metrics import mean_squared_error, r2_score
import pandas as pd
import statsmodels.api as sm
import seaborn as sns
sns.set()
data=pd.read_csv("CO2_GDP_USA_Data.csv")
data.head()
plt.rcParams["font.size"] = "20"
y = np.array(data["co2 per capita (metric tons)"])
x = np.array(data["gdp per capita (USD)"]).reshape((-1, 1))
plt.scatter(x , y , marker = "+" , color = "blue")
plt.ylabel("CO$_2$ Emissions Per Capita (Tons)")
plt.xlabel("GDP Per Capita (USD)")
regr = linear_model.LinearRegression()
```

```
                        OLS Regression Results
===============================================================================
Dep. Variable:                      y   R-squared (uncentered):             0.924
Model:                            OLS   Adj. R-squared (uncentered):        0.920
Method:                 Least Squares   F-statistic:                        243.7
Date:                Sun, 17 Apr 2022   Prob (F-statistic):              1.15e-12
Time:                        08:05:32   Log-Likelihood:                   -64.030
No. Observations:                  21   AIC:                                130.1
Df Residuals:                      20   BIC:                                131.1
Df Model:                           1
Covariance Type:            nonrobust
===============================================================================
                 coef    std err          t      P>|t|      [0.025      0.975]
-------------------------------------------------------------------------------
x1             0.0004   2.25e-05     15.611      0.000       0.000       0.000
===============================================================================
Omnibus:                        3.054   Durbin-Watson:                      0.030
Prob(Omnibus):                  0.217   Jarque-Bera (JB):                   1.289
Skew:                          -0.113   Prob(JB):                           0.525
Kurtosis:                       1.808   Cond. No.                           1.00
===============================================================================

Warnings:
[1] Standard Errors assume that the covariance matrix of the errors is correctly specified.
```

Figure 15.2 OLS regression results from Program_15a.py. The details of the results can be investigated by reading the notes from the statsmodels URL.

```python
regr.fit(x , y)
y_pred = regr.predict(x)
print("Gradient: \n", regr.coef_)
print("y-Intercept: \n", regr.intercept_)
print("MSE: %.2f"% mean_squared_error(y , y_pred))
print("R2 Score: %.2f" % r2_score(y , y_pred))
plt.plot(x , y_pred , color = "red")
plt.show()
sm.add_constant(x)
results = sm.OLS(y , x).fit()
print(results.summary())
```

Statistical Tests used in Figure 15.2

Omnibus test: Detects any of a broad range of departures from a specific null hypothesis.

Skew: The degree of asymmetry observed in a probability distribution.

Kurtosis: The degree to which data clusters in the tails or the peak of a frequency distribution.

Durbin-Watson: The autocorrelation in the residuals.

Jarque-Bera: A goodness-of-fit test to check whether the data have the kurtosis and skewness close to a normal distribution.

15.2   MARKOV CHAINS

A Markov chain is a stochastic process that describes a sequence of possible events where the probability of the next event is only dependent on the current event, and not how it got there. Future and past events are independent, and only conditional on the present event. Markov chains are extremely useful in data science and have a wide range of applications in ranking of websites in search engines, natural language processing (NLP) algorithms, marketing, financial stock price movement, Brownian motion in physics, information theory and computer performance evaluation, to name but a few. Readers are directed to [11], which covers stochastic stability, and [13] for more theory and applications. Readers may also be interested in hidden Markov chain models [5], where one may predict a sequence of unknown (hidden) events from a set of observed events. This section is restricted to discrete-time stochastic processes. A Markov chain can be represented by a directed graph, an example is illustrated in Figure 15.3, and a corresponding state transition matrix, T say, listed as equation (15.3), in this case. The ijth element of T represents the conditional probability of the chain in state j given that it was in state i in the previous time interval. In general, a discrete-time Markov chain is a sequence of random variables $X_1, X_2, \ldots, X_n$, giving a transition probability:

$$P\left(X_{n+1} = x | X_1 = x_1, X_2 = x_2, \ldots, X_n = x_n\right) = P\left(X_{n+1} = x | X_n = x_n\right),$$

where the conditional probabilities are well defined, and $P\left(X_1 = x_1, \ldots X_n = x_n\right) > 0$. To describe a Markov chain, an initial state distribution and transition probabilities

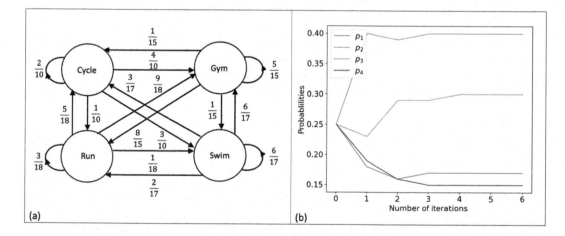

Figure 15.3 (a) Markov chain directed graph, where circles represent workout events and the directed edges are probability transitions. (b) Convergence of the probability vector to the steady state vector, $\pi = [0.17, 0.4, 0.3, 0.15]$, after five iterations, where p_i are probabilities.

Table 15.1 Exercise logs from the last 60 workout sessions.

Set	Cycle	Gym	Run	Swim	Total
Cycle	2	4	1	3	10
Gym	1	5	8	1	15
Run	5	9	3	1	18
Swim	3	6	2	6	17
Totals	11	24	14	11	60

are required. Future states are computed using recursion, and can be used to calculate a stationary distribution, thus

$$v_{i+1} = v_i \times T,$$

where v_i is a probability vector on the ith iterate, so, given an initial probability distribution, v_0 say:

$$v_1 = v_0 T, \quad v_2 = v_1 T = (v_0 T) T = v_0 T^2, \quad \dots, v_n = v_0 T^n.$$

The stationary distribution, usually referred to as π, is defined by

$$\pi = \lim_{n \to \infty} v_0 T^n.$$

Definition 15.2.1. A Markov chain is called regular if all powers of the transition matrix T have only positive non-zero elements. All regular Markov chains converge over time [17].

Example 15.2.1. An individual keeps a log of their workout transitions, cycling, going to the gym, running and swimming. Table 15.1 displays the results from the last 60 workouts. Plot a directed graph to represent the Markov chain, write down the transition matrix. Given that the starting probability vector is $v_0 = [0.25, 0.25, 0.25, 0.25]$, determine the steady state, and plot a graph showing convergence to the steady state.

Solution. From Table 15.1, the transition probabilities are easily calculated and displayed in Table 15.2. The corresponding transition matrix is given as an equation (15.3). Figure 15.3(a) shows the Markov chain diagram for the workout model of an individual who regularly cycles, goes to the gym runs and swims. Program_15b.py is used to plot Figure 15.3(b) and also lists the probability vectors v_0 to v_5, where the system has reached a steady state. The probability vector π is called the steady state vector of the Markov chain, and indicates the probabilities of being in the various states as time gets large.

Table 15.2 Transition probabilities for each sequence of workout sets.

Set	Cycle	Gym	Run	Swim	Total
Cycle	$\frac{2}{10}$	$\frac{4}{10}$	$\frac{1}{10}$	$\frac{3}{10}$	1
Gym	$\frac{1}{15}$	$\frac{5}{15}$	$\frac{8}{15}$	$\frac{1}{15}$	1
Run	$\frac{5}{18}$	$\frac{9}{18}$	$\frac{3}{18}$	$\frac{1}{18}$	1
Swim	$\frac{3}{17}$	$\frac{6}{17}$	$\frac{2}{17}$	$\frac{6}{17}$	1

The transition matrix corresponding to Table 15.2 is given by:

$$T = \begin{pmatrix} \frac{2}{10} & \frac{4}{10} & \frac{1}{10} & \frac{3}{10} \\ \frac{1}{15} & \frac{5}{15} & \frac{8}{15} & \frac{1}{15} \\ \frac{5}{18} & \frac{9}{18} & \frac{3}{18} & \frac{1}{18} \\ \frac{3}{17} & \frac{6}{17} & \frac{2}{17} & \frac{6}{17} \end{pmatrix}. \tag{15.3}$$

```python
# Program_15b.py: Markov Chain.
import numpy as np
import matplotlib.pyplot as plt
T = np.array([[2/10,4/10,1/10,3/10], \
              [1/15,5/15,8/15,1/15], \
              [5/18,9/18,3/18,1/18], \
              [3/17,6/17,2/17,6/17]
              ])
n = 20
v=np.array([[0.25, 0.25, 0.25 , 0.25]])
print(v)
vHist = v
for x in range(n):
  v = np.dot(v , T).round(2)
  vHist = np.append(vHist , v , axis = 0)
  if np.array_equal(vHist[x] , vHist[x-1]):
      print("Steady state after" , x , "iterations.")
      break
  print(v)
plt.rcParams["font.size"] = "14"
plt.xlabel("Number of iterations")
plt.ylabel("Probablilities")
plt.plot(vHist)
#plt.rcParams["linecolor"]
plt.legend(["$p_1$","$p_2$","$p_3$","$p_4$"],loc="best")
plt.show()
```

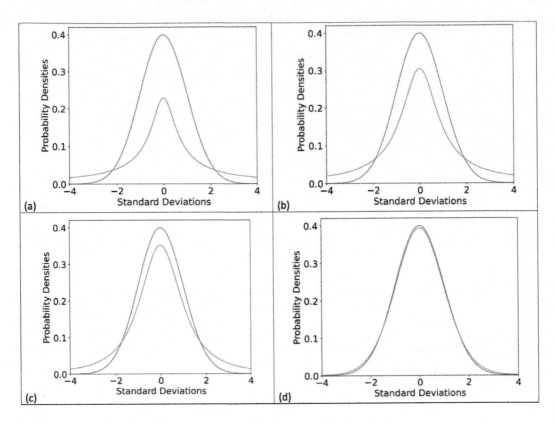

Figure 15.4 (a)–(d) Stills of the animation of a student-t curve (orange curve) produced using Program_15c.py. The curve approaches a normal probability distribution curve (blue curve) as the degrees of freedom, df say, tends to infinity.

15.3 THE STUDENT T-TEST

A student's t-test, or t-test, is a statistical hypothesis test where the test statistic follows a distribution known as the student's t-distribution (or t-distribution) under the null hypothesis. Figure 15.4 shows how the t-distribution approximates the normal distribution as the degrees of freedom goes to infinity. Program_15c.py produces an animation, where the degrees of freedom goes up to 25. The Python commands from the scipy library (where scipy.stats is imported as stats) for plotting a t-distribution and a normal distribution are given by **stats.t.pdf(x,df)** and **stats.norm.pdf(x)**, respectively, where x is a real number and df is the degrees of freedom.

```
# Program_15c.py: Animation of a Student-T Curve.
import numpy as np
import matplotlib.pyplot as plt
import matplotlib.animation as animation
import scipy.stats as stats
fig, ax = plt.subplots()
x = np.arange(-4 , 4 , 0.01)
```

```
y1 = stats.norm.pdf(x)
y2 = 0 * x
plt.plot(x , y1)
ax = plt.axis([-4 , 4 , 0 , 0.42])
Student_T, = plt.plot(x, y2) # Base line.
def animate(n):
    Student_T.set_data(x, stats.t.pdf(x, n))
    return Student_T,
# Create animation.
myAnimation = animation.FuncAnimation(fig, animate,\
    frames=np.arange(0, 25, 0.1), interval=1, blit=True, repeat=True)
plt.ylabel("Probability Densities")
plt.xlabel("Standard Deviations")
plt.rcParams["font.size"] = "20"
plt.show()
```

There are three types of t-test, a one sample t-test, a two sample t-test and a paired t-test. This section will consider an example of a two sample t-test, where the dependent variables are normally distributed in each group and there is homogeneity of variance, or the variances are almost equal. If these conditions do not hold, then other tests will be required. There are many statistical tests available, see [9] and [10], each lists one hundred tests.

Definition 15.3.1. A t-test for two population means, where the variances are equal but unknown, has the hypotheses and test statistic:

Hypothesis and Alternative Hypothesis:

$$1. H_0 : \mu_1 - \mu_2 = \mu_0, \quad H_1 : \mu_1 - \mu_2 \neq \mu_0;$$

$$2. H_0 : \mu_1 - \mu_2 = \mu_0, \quad H_1 : \mu_1 - \mu_2 > \mu_0.$$

Test statistic:

$$t = \frac{\bar{x}_1 - \bar{x}_2 - \mu_0}{s_p\sqrt{\frac{1}{n_1} + \frac{1}{n_2}}},$$

where

$$s_p^2 = \frac{(n_1 - 1)s_1^2 + (n_2 - 1)s_2^2}{n_1 + n_2 - 2},$$

where $\bar{x}_1, \bar{x}_2, s_1, s_2, n_1, n_2$, are the observed means, standard deviations and sample sizes for the two data sets, respectively, and s_p is the pooled standard deviation.

Example 15.3.1. A lecturer measures the height of female and male students in a class (in inches) and obtains the following lists: Female heights: [58.9, 65.2, 63.4,64.5, 61.7, 66.0, 62.9, 65.6, 61.9, 63.7, 68.1, 61.8, 63.4, 58.9, 58.4, 60.8, 70.1, 62.3, 61.7, 63.1, 62.3, 61.8, 66.3, 62.0, 65.3, 61.2, 67.0, 60.2, 63.9, 63.4, 63.7, 60.3, 63.4, 60.6, 67.4, 66.9, 60.9, 63.6, 59.6, 59.7, 65.0], and Male heights: [73.8, 68.8, 74.1, 71.7, 69.9,

67.3, 68.8, 68.3, 67.3, 68.8, 68.3, 67.0, 63.5, 71.2, 71.6, 64.8, 69.3, 69.2, 67.6, 72.4, 64.3, 61.5, 74.8, 67.8, 69.6, 60.7, 71.3, 70.4, 67.0, 71.5, 72.7, 69.1, 69.0, 71.6, 70.4]. Plot box and whisker plots, histograms, and quantile-quantile plots of the data, and carry out a t-test on the data. What can you conclude?

Solution. Program_15d.py lists the commands for plotting box and whisker plots, histogram plots and quantile-quantile plots for the data. The results show that both samples have almost normal distribution. The results are displayed in Figure 15.5.

The Levene test is used to test equivalence of variances. The Levene test (stats.levene in the program) gives the result:

LeveneResult(statistic=0.40649312655884273, pvalue=0.5257226981096292).

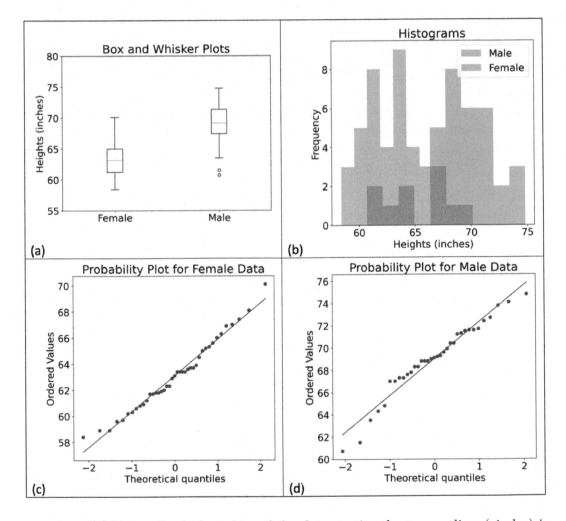

Figure 15.5 (a) Box and whisker plots of the data, notice the two outliers (circles) in the male data. (b) Histograms of the data. (c) and (d) Quantile-quantile (or Q-Q) plots, both distributions are nearly normal.

The p-value of the test is greater than the threshold value of 0.05, we accept the null-hypothesis that, on average, the data sets have almost the same variance.

The t-test is used to test the differences between two means. The null-hypothesis in this case is: females and males in the class are equally tall, on average. The t-test (stats.ttest in the program) gives the result:

Ttest_indResult(statistic=-8.63262625399211, pvalue=8.229888352642539e-13),

and since the p-value is less than the threshold value of 0.05, we reject the null-hypothesis of no difference between female and male heights, and conclude that, on average, males are taller than females.

```
# Program_15d.py: Male and Female Sample Heights.
import matplotlib.pyplot as plt
import pylab
import scipy.stats as stats
FD = [58.9, 65.2, 63.4 , 64.5, 61.7, 66.0, 62.9, 65.6, \
        61.9, 63.7, 68.1, 61.8, 63.4, 58.9, 58.4, 60.8, \
      70.1, 62.3, 61.7, 63.1, 62.3, 61.8, 66.3, \
      62.0, 65.3, 61.2, 67.0, 60.2, 63.9, 63.4, \
      63.7, 60.3, 63.4, 60.6, 67.4, 66.9, 60.9, 63.6, \
        59.6, 59.7, 65.0]
MD = [73.8, 68.8, 74.1, 71.7, 69.9, 67.3, 68.8, 68.3, \
        67.3, 68.8, 68.3, 67.0, 63.5, 71.2, 71.6, 64.8, \
      69.3, 69.2, 67.6, 72.4, 64.3, 61.5, 74.8, 67.8,\
      69.6, 60.7, 71.3, 70.4, 67.0, 71.5, 72.7, 69.1, \
        69.0, 71.6, 70.4]
data = [FD , MD]
fig , ax = plt.subplots()
ax.set_ylim([55, 80])
ax.boxplot(data)
plt.xticks([1, 2], ["Female", "Male"])
plt.ylabel("Heights (inches)")
plt.title("Box and Whisker Plots")
plt.show()
plt.figure()
plt.hist(MD, label="Male" , alpha=0.5)
plt.hist(FD, label="Female" , alpha=0.5)
plt.legend()
plt.title("Histograms")
plt.xlabel("Heights (inches)")
plt.ylabel("Frequency")
plt.figure()
stats.probplot(MD, dist="norm", plot=pylab)
```

```
pylab.title("Probability Plot for Male Data")
pylab.show()
plt.figure()
stats.probplot(FD, dist="norm", plot=pylab)
pylab.title("Probability Plot for Female Data")
pylab.show()
# Use Levene's test to test for equality of variances.
print(stats.levene(FD,MD,center= "mean"))
# Student t-test.
print(stats.ttest_ind(FD,MD))
```

15.4   MONTE-CARLO SIMULATION

The formal foundations of Monte-Carlo simulation were set by John Louis von Neumann and Stanislaw Ulam in the 1940s, using random experiments to solve deterministic problems. The theory of probability density functions, inverse cumulative distribution functions and random number generators was developed using digital computers. The name is derived from the popular gambling destination in Monaco, where games like roulette, dice, cards and slot machines generate random outcomes which are central to the simulation technique. This section will present a simple example in Roulette, and it is shown why the house always wins. There are a vast range of applications of Monte-Carlo simulations throughout the sciences and beyond, for example, it can be used to estimate π, solve integrals, and solve problems in finance, project management, manufacturing, engineering, insurance, transportation and even nuclear reactions. For more information on the theory and specific applications, the reader is directed to [1], [2] and [14], and for a Python book, see [4].

In European roulette, the wheel consist of 37 pockets, numbered zero through to 36. The number zero is colored green and the other 36 numbers are colored red and black, where 18 are colored red and 18 are colored black. There are various ways to gamble in roulette, for those interested, I recommend [16], however, I strongly advise against gambling, and readers should be aware that the house always wins in the end, as we show here. As a simple example, a gambler is going to bet on red, the roulette croupier spins the wheel, and throws the ball in the opposite direction to the spin of the wheel. Eventually, the ball settles in to one of the pockets. If the ball lands in a red pocket, then the gambler wins and receives prize money, however, if the ball lands in a black or green pocket, then the house wins. Simple statistics reveals that the chance of winning is $\frac{18}{37}$, and the chance of losing is $\frac{19}{37}$. The Monte Carlo simulations show quite clearly that in the end the house always wins.

Example 15.4.1. A gambler enters a casino with a bankroll of 10,000 euros, and places 100 euro bets on the ball landing on red every time. If the gambler loses on a spin, then they lose the wager, and if they win, then they get a profit of the wager. Run a simple Monte Carlo simulation for Roulette, where the gambler bets on red on every spin. What can you conclude?

Solution. Program_15e.py lists the Python code for the Monte Carlo simulations. Figure 15.6(a) shows the result of 100 simulations of 3,700 spins, and the bankrolls against the number of spins of the roulette wheel. The number 3,700 is chosen, as this is the average number of spins required for the gambler to lose all of their bankroll. Figure 15.6(b) shows 1000 simulations of 1850 spins of the roulette wheel. There is clear convergence to the average bankroll reaching 5000 euros. Can you explain why?

```python
# Program_15e.py: Monte Carlo Simulation. European roulette.
# A gambler keeps betting on the ball landing on a red pocket.
import random
import matplotlib.pyplot as plt
def spin():
    pocket = random.randint(1,37)
    return pocket
def roulette(bankroll, wager, spins):
    Num_Spins = []
    balance = []
# The first spin.
    num_spins = 1
# Set maximum number of spins.
    while num_spins < spins:
        # Payout if ball lands on red.
        if (spin() % 2) == 0:
            # Add money to bankroll.
            bankroll = bankroll + wager
            # Increase number of spins.
            Num_Spins.append(num_spins)
            # Append the balance.
            balance.append(bankroll)
        # Gambler loses if ball lands in a black or green pocket.
        else:
            # Subtract the money from the bankroll.
            bankroll = bankroll - wager
            # Increase the number of spins.
            Num_Spins.append(num_spins)
            # Append the balance.
            balance.append(bankroll)
        num_spins = num_spins + 1
    plt.plot(Num_Spins , balance)
    final_balances.append(balance[-1])
    return(final_balances)
simulations = 1
#Create a list for calculating final funds.
final_balances = []
```

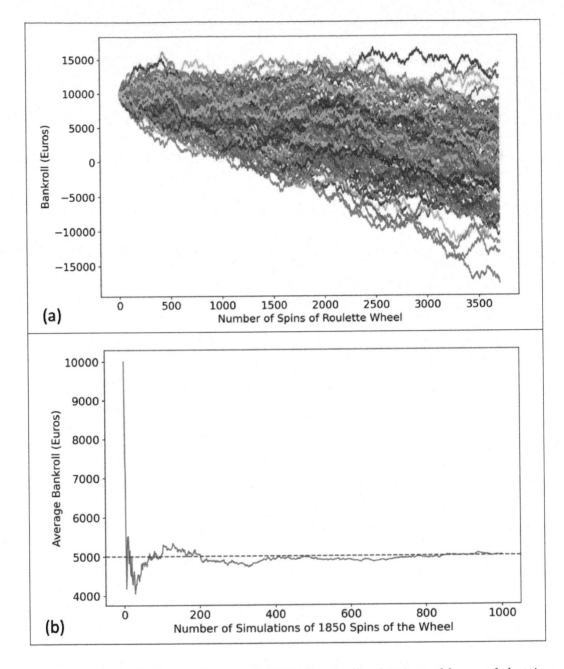

Figure 15.6 Monte Carlo simulations: (a) The bankrolls of 100 gamblers each betting on 3700 spins, where the gambler wages 100 euros on each spin. In this case, the program shows that on average, the gambler would leave the casino with a small bankroll, or even go in to debt. Each time you run the simulation, you will get different answers. (b) As the number of simulations of 1850 spins of the roulette wheel goes large, one can clearly see that the average bankroll tends to 5000 euros. So on average, the gambler would expect to lose half of the bankroll after 1850 spins.

```
while simulations <= 100:
    end_balance = roulette(10000 , 100 , 3700)
    simulations = simulations + 1
plt.rcParams["font.size"] = "20"
plt.xlabel("Number of Spins of Roulette Wheel")
plt.ylabel("Bankroll (Euros)")
plt.show()
print("The player starts the game with 10,000 euros and ends with " \
+ str(sum(end_balance)/len(end_balance)), "euros")
```

EXERCISES

15.1 Find the data on the internet for carbon dioxide emissions per capita, and gross domestic product per capita, for USA, in the years 1960 to 2000, and perform a linear regression on that data. What can you conclude?

15.2 In finance, a bull market is where stock prices are rising, economic production is strong, inflation is low, and the employment rate is high. In contrast, a bear market is where stock prices are falling, the economy slows, and inflation and unemployment rise. A stagnant market is one in which the market neither grows nor shrinks. A simple transition matrix for a bull-bear-stagnant market system is given by:

$$T = \begin{pmatrix} 0.9 & 0.075 & 0.025 \\ 0.15 & 0.8 & 0.05 \\ 0.25 & 0.25 & 0.5 \end{pmatrix}.$$

Plot a Markov chain directed graph for this system and determine the steady state. Use the initial state vector $v_0 = [0.1, 0.1, 0.8]$, and plot the probabilities on a graph until the system converges to the steady state.

15.3 Analysis of variance (ANOVA) generalizes the student t-test and provides a test of whether two or more population means are equal. As with the t-test, certain conditions have to be met in order for the test results to be reliable. A one-way ANOVA has one independent variable. A pharmacology research laboratory monitors the concentrations of nitric oxide (NO) in rat plasma after a certain drug is administered, see Table 15.3. Research ANOVA on the web or read [8], and use Python to conduct a one-way ANOVA test on the data. Use scipy.stats.f_oneway and statsmodels to determine whether there is a significant change in the concentration of NO in rat plasma.

15.4 Run a Monte-Carlo simulation for the situation described in Example 15.4.1, to obtain a plot similar to Figure 15.6(b). What can you conclude?

Table 15.3 Amount of nitrous oxide in rat plasma $\mu moll^{-1}$ after a drug is administered.

Control (Placebo)	Drug A	Drug B	Drug C	Drug D
83	21	52	49	13
90	63	26	66	69
50	66	77	80	22
75	50	56	14	38
112	54	44	102	55
49	89	89	41	42
139	66	104	83	84
60	89	41	89	44
101	85	89	71	43
81	40	41	89	31

Solutions to the Exercises may be found here:

`https://drstephenlynch.github.io/webpages/Solutions_Section_2.html`.

FURTHER READING

[1] Barbu, A. (2020). *Monte Carlo Methods.* Springer, New York.

[2] Benton, D.J. (2018). *Monte Carlo Simulation: The Art of Random Process Characterization.* Independently published.

[3] Bruce, P., Bruce, A. and Gedeck, P. (2020). *Practical Statistics for Data Scientists: 50+ Essential Concepts Using R and Python 2nd Ed.* O'Reilly Media, Sebastopol, CA.

[4] Ciaburro, G. (2020). *Hands-On Simulation Modeling with Python: Develop simulation models to get accurate results and enhance decision-making processes.* Packt Publishing, Birmingham.

[5] Coelho, J.P., Pinho, T.M. and Boaventura-Cunha, J. (2021). *Hidden Markov Models: Theory and Implementation using MATLAB.* CRC Press, Boca Raton, FL.

[6] Frost, J. (2020). *Regression Analysis: An Intuitive Guide for using and Interpreting Linear Models.* Statistics by Jim Publishing.

[7] Haslwanter, T. (2018). *An Introduction to Statistics with Python: With Applications in the Life Sciences.* Springer, New York.

[8] Judd, C.M. McClelland, G.H. and Ryan, C.S. (2017). *Data Analysis: A Model Comparison Approach To Regression, ANOVA, and Beyond, 3rd Ed.* Routledge, Oxford.

[9] Kanji, G.K. (2006). *100 Statistical Tests, 3rd Ed.* Independently Published.

[10] Lewis, N.D. (2020). *100 Statistical Tests in R*. SGE Publications Ltd, Thousand Oaks, CA.

[11] Meyn, S. and Tweedie, R.L. (2009). *Markov Chains and Stochastic Stability, 2nd Ed.* Cambridge University Press, Cambridge.

[12] Montgomery, D.C., Peck, E.A. and Vining, G.G. (2012). *Introduction to Linear Regression Analysis, 5th Ed.* Wiley, New York.

[13] Privault, N. (2018). *Understanding Markov Chains: Examples and Applications.* Springer, New York.

[14] Runinstein, R.Y. (2016). *Simulation and the Monte Carlo Method, 3rd Ed.* Wiley, New York.

[15] Saha, A. (2015). *Doing Math with Python: Use Programming to Explore Algebra, Statistics, Calculus, and More!* No Starch Press, San Francisco, CA.

[16] Sander, D. (2017). *How To Play Roulette: The Ultimate Guide to Roulette Strategy, Rules and Roulette Systems for Greater Winnings.* CreateSpace Independent Publishing Platform, Scotts Valley, CA.

[17] Sekhon, R. and Bloom, R. (2020). *Applied Finite Mathematics.* LibreTexts, Columbus, OH.

III

Artificial Intelligence

## CHAPTER 16

Brain Inspired Computing

Brain inspired computing is an invention by the author and his colleague, Jon Borresen. We applied for our first patent in 2010, which was granted in 2012 [15], and a second patent was granted in 2015 [16]. The idea is simple to describe, and the author has been delivering inspirational talks on the subject to local school children aged 11–18 since 2012, as part of Schools Liaison, Speakers for Schools and Widening Participation. The chapter is all about networks of coupled oscillators, for more general information, the reader is directed to the recent book by Rico Berner [1]. The chapter starts with a description of the complicated Hodgkin-Huxley model of a biological neuron, and introduces a simplified version known as the Fitzhugh-Nagumo (FN) model. In the next section, using the FN model, the binary oscillator half-adder is described. The half-adder is the basic component of all conventional logic computers. Next, an oscillatory set reset flip-flop is described, which is used to store memory in computers, and has many other applications. The chapter ends with a description of the potential applications of the invention, and the progress of building commercial devices to-date.

The reader should note that brain inspired computing—using brain dynamics to build conventional computers, is the opposite of artificial intelligence (AI)—using conventional computers to simulate brain dynamics, however, understanding how biological neurons work is fundamentally important in AI.

Aims and Objectives:

- To provide a brief history of modeling neurons.

- To review basic operations of neurons, including excitation and inhibition.

- To introduce threshold oscillator logic and memory.

At the end of the chapter, the reader should be able to

- model biological neurons;

- perform simple binary logic and memory operations using threshold oscillators;

- plot time series data to illustrate functionality;

- appreciate the potential applications of the invention.

DOI: 10.1201/9781003285816-16

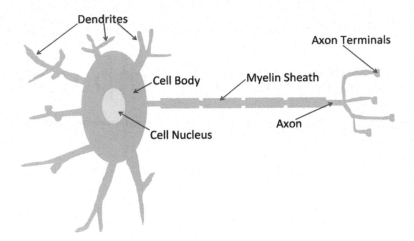

Figure 16.1 Schematic of a neuron modeled by Hodgkin and Huxley. This structure is important when constructing mathematical models of neurons and neural networks in the next chapter.

16.1 THE HODGKIN-HUXLEY MODEL

In 1952, Hodgkin and Huxley [13] were investigating the action potentials in a giant squid axon. A schematic of a neuron is presented in Figure 16.1, and consists of a cell body, (or soma) which contains the cell nucleus comprising genetic material in the form of chromosomes; dendrites, which receive signals from connected pre-synaptic neurons; the axon surrounded by a myellin sheath for insulation, and finally the axon terminals, which transmit signals to connected post-synaptic neurons. Biological neurons are natural threshold oscillators, which means that they will only start to oscillate once a certain threshold voltage is reached. Hodgkin and Huxley simulated the biological processes using simple electrical circuits and derived the following system of four ODEs based on those circuits:

$$C\frac{dV}{dt} = I - g_{Na}m^3h\left(V - V_{Na}\right) - g_K n^4\left(V - V_K\right) - g_L\left(V - V_L\right),$$
$$\frac{dm}{dt} = \alpha_m(1 - m) - \beta_m m$$
$$\frac{dh}{dt} = \alpha_h(1 - h) - \beta_h h$$
$$\frac{dn}{dt} = \alpha_n(1 - n) - \beta_n n, \tag{16.1}$$

where V is membrane potential, V_{Na}, V_K, V_L, C and g_L are all constants determined from experimental data, and g_{Na} and g_K are both functions of time and membrane potential. The three dimensionless quantities m, h, and n represent sodium, potassium and leakage gating variables, respectively, and α_i and β_i are the transition rate constants for the i-th ion channel.

Example 16.1.1. Based on experimental data, the following parameter values were chosen:

$$\alpha_m = \frac{0.1(V + 40)}{1 - \exp(-0.1(V + 40))}, \quad \beta_m = 4\exp(-0.0556(V + 65)),$$

$$\alpha_h = 0.07\exp(-0.05(V + 65)), \quad \beta_h = \frac{1}{1 + \exp(-0.1(V + 35))},$$

$$\alpha_n = \frac{0.01(V + 55)}{1 - exp(-0.1(V + 55))}, \quad \beta_n = 0.125\exp(-0.0125(V + 65)), \quad (16.2)$$

and additionally,

$$C = 1\,\mu\mathrm{Fcm}^{-2},$$

$$g_L = 0.3\,\mathrm{mmhocm}^{-2}, g_K = 36\,\mathrm{mmhocm}^{-2}, g_{Na} = 120\,\mathrm{mmhocm}^{-2},$$

$$V_L = -54.402\,\mathrm{mV}, V_K = -77\,\mathrm{mV}, V_{Na} = 50\,\mathrm{mV}. \quad (16.3)$$

Use Python to plot neuron action potentials.

Solution. The Python program for solving the ODEs (16.1) using equations (16.2) and (16.3) is listed as Program_16a.py. The plots of the neuron action potentials are displayed in Figure 16.2. As the input current increases, the frequency of oscillation increases up to some limit, where the neuron stops oscillating again. The reader is encouraged to run the program for differing input currents to see how it affects the model. For the parameters chosen here, the threshold current is $IC = 6.23$mA.

```python
# Program_16a.py: The Hodgkin-Huxley Equations.
import numpy as np
import matplotlib.pyplot as plt
from scipy.integrate import odeint
C_m,g_Na,g_K,g_L,V_Na,V_K,V_L=1,120,36,0.3,50,-77,-54.402

def alpha_m(V): return 0.1*(V + 40.0) / (1.0-np.exp(-0.1*(V+40.0)))
def beta_m(V): return 4.0 * np.exp(-0.0556 * (V + 65.0))
def alpha_h(V): return 0.07 * np.exp(-0.05 * (V + 65.0))
def beta_h(V): return 1.0 / (1.0 + np.exp(-0.1 * (V + 35.0)))
def alpha_n(V): return 0.01*(V + 55.0) / (1.0-np.exp(-0.1*(V+55.0)))
def beta_n(V): return 0.125 * np.exp(-0.0125 * (V + 65))

def I_Na(V,m,h): return g_Na * m**3 * h * (V - V_Na)
def I_K(V, n): return g_K * n**4 * (V - V_K)
def I_L(V): return g_L * (V - V_L)

# Input current
thresh = 6.23    # Threshold current.
```

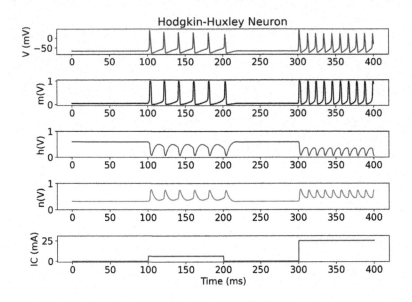

Figure 16.2 Numerical solutions of equations (16.1) to (16.3). The upper blue curve is the neuron action potential. the middle black, red and green curves are gating variables m, h, and n, respectively, and the lower magenta curve shows input current, IC. For $100 \leq t \leq 200$ ms, the input current is set at a threshold value of $IC = 6.23$mA.

```
def Input_current(t): return thresh*(t>100)-thresh*(t>200)+25*(t>300)

t = np.arange(0.0, 400.0, 0.1)
# Set up the ODEs, see equations (21.3)
def hodgkin_huxley(X, t):
    V, m, h, n = X
    dVdt = (Input_current(t)-I_Na(V,m,h)-I_K(V,n)-I_L(V))/C_m
    dmdt = alpha_m(V) * (1.0 - m) - beta_m(V) * m
    dhdt = alpha_h(V) * (1.0 - h) - beta_h(V) * h
    dndt = alpha_n(V) * (1.0 - n) - beta_n(V) * n
    return (dVdt, dmdt, dhdt, dndt)

y0 = [-65, 0.05, 0.6, 0.32]
X = odeint(hodgkin_huxley, y0, t)
V , m , h , n = X[:,0] , X[:,1] , X[:,2] , X[:,3]
ina , ik , il = I_Na(V,m,h) , I_K(V,n) , I_L(V)
plt.subplots_adjust(hspace = 1)
plt.figure(1)
plt.subplot(5, 1, 1)
plt.title("Hodgkin-Huxley Neuron")
plt.plot(t, V, "b")
```

```
plt.ylabel("V (mV)")
plt.subplot(5, 1, 2)
plt.plot(t, m, "k")
plt.ylabel("m(V)")
plt.subplot(5, 1, 3)
plt.plot(t, h, "r")
plt.ylim(0, 1)
plt.ylabel("h(V)")
plt.subplot(5, 1, 4)
plt.plot(t, n, "g")
plt.ylim(0, 1)
plt.ylabel("n(V)")
plt.subplot(5, 1, 5)
plt.plot(t, Input_current(t), "m")
plt.ylabel("IC (mA)")
plt.xlabel("Time (ms)")
plt.ylim(-1, 31)
plt.show()
```

In 1961 and 1970, Fitzhugh [10] and Nagumo [19] were able to simplify the Hodgkin-Huxley equations to model the action potential of a spiking neuron, the two ODEs used to model a single neuron are written as:

$$\frac{du}{dt} = i + u(u - \theta)(1 - u) - v, \quad \frac{dv}{dt} = \epsilon(u - \gamma v), \qquad (16.4)$$

where u is a fast variable (in biological terms—the action potential) and v represents a slow variable (biologically—the sodium gating variable). The parameters θ, γ and ϵ dictate the threshold, oscillatory frequency and the location of the critical points for u and v, respectively. A neuron will begin to oscillate when the input current i is above a critical threshold, i_T, say. Equations (16.4) are known as the Fitzhugh-Nagumo (FN) equations and will be used in the next two sections of the chapter when simulating logic and memory devices.

16.2   THE BINARY OSCILLATOR HALF-ADDER

The basic idea behind the binary oscillator half-adder is very simple and involves switching threshold oscillators (neurons) on using excitation, and switching oscillators off using inhibition. Chemical excitation and inhibition in biological neurons was described by Destexhe et al. [7] in 1994. Figure 16.3 shows the results of the mathematical model of chemical excitation and inhibition. Put simply, an excitatory neuron can switch a postsynaptic neuron on (as long as the threshold is reached), and an inhibitory neuron can switch a postsynaptic neuron off. How excitation and inhibition works will depend on the technology being adopted.

For a general introduction to the basis of physical computation, I recommend Feynman's book [9], and for an introduction to digital design and computer architecture, see [12]. Bucaro [4] provides an easy free open source logic gate simulator. The

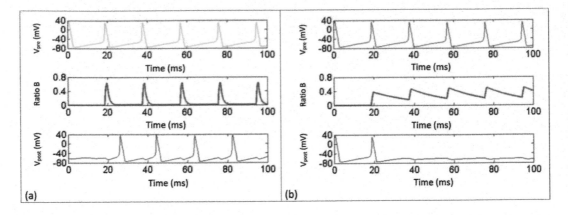

Figure 16.3 Mathematical modeling of chemical excitation and inhibition. (a) Excitation, the upper green spikes are from the presynaptic neuron, the magenta trace in the middle depicts the ratio of bound receptors, and the lower blue spikes show the action potential in the postsynaptic neuron. (b) Inhibition, the upper red spikes are from the presynaptic neuron, the magenta trace in the middle depicts the ratio of bound receptors, and the lower blue trace shows the action potential in the postsynaptic neuron, which is turned off.

binary half-adder shown in Figure 16.4(a) consists of an XOR gate which requires five transistors and an AND gate which can be built with two transistors. Neural networks of these logic gates are discussed in the next chapter. The truth table for the binary half-adder is shown in Figure 16.4(b), where A and B are binary inputs, zeros and ones, and S is the sum while C is the carry. The binary half-adder is used to add two bits together in binary. Thus, $0 + 0 = 00$, see column 1, the sum is zero and the carry is zero. Now, $0 + 1 = 01$, see column two, $1 + 0 = 01$, see column three, and finally, $1 + 1 = 10$, in binary, as indicated in column four. Think of S giving the units and C giving the tens in binary.

Figure 16.4(c) shows the schematic of our invention, it is called a binary oscillator half-adder. It consists of two inputs A and B which are either oscillating or not. The green arrows are excitatory synaptic connections to threshold oscillators labelled O1 and O2. O1 has a lower threshold than O2 or the synaptic weights can differ, as seen in Example 16.2.1. The red arrow depicts an inhibitory synaptic connection from O2 to O1, and S is the sum and C is the carry. I will now explain how the invention works. Looking down the columns in Figure 16.4(c). Consider column one, if A and B are not oscillating, then there is no input to O1 and O2, and they both remain off. Next consider columns two and three. If either A or B is oscillating, then the device is set such that sufficient current enters O1, threshold is exceeded and O1 starts to oscillate, however, there is insufficient current reaching O2, threshold is not reached, and O2 does not oscillate. Finally, consider column four. If A and B are both oscillating, then the thresholds of O1 and O2 are reached and they both start to oscillate, however, O2 immediately inhibits O1, and switches it off. Looking at Figures 16.4(b) and (d), the reader can see that the binary oscillator half-adder is behaving in the

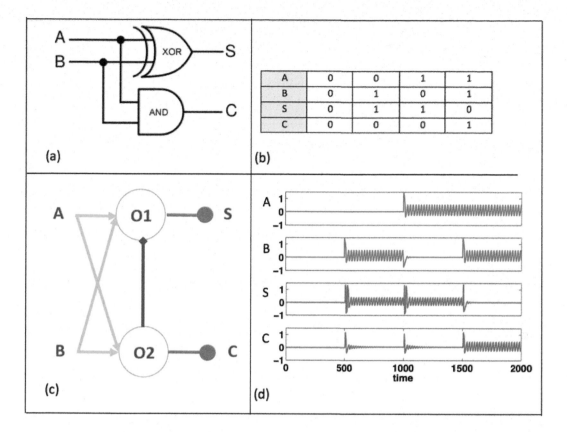

Figure 16.4 Binary half-adders. A and B are inputs, S is the sum and C is the carry. (a) A binary half-adder made of an XOR and AND logic gates. (b) Truth table for a binary half-adder. (c) Binary oscillator half-adder. The green arrows represent excitatory connections and the red arrow depicts an inhibitory connection. O1 and O2 are threshold oscillators, where O1 and O2 have the same threshold but the synaptic weights differ. (d) Output for the binary oscillator half-adder. An oscillation is equivalent to a binary one and no oscillation is zero. Compare with the truth table in (b).

correct way. For further detail on how excitation and inhibition are simulated, and examples of a binary oscillator full-adder (used to add three bits together), a seven bit adder, and a 2×2 bit multiplier, see our PLoS ONE paper [2].

Example 16.2.1. Using the FN equations (16.4), simulate the binary oscillator half-adder as depicted in Figure 16.4(c), using a sigmoidal transfer function given by $\sigma(x) = \frac{1}{1+\exp(m(c-x))}$, where c and m are constants. Transfer functions are discussed in more detail in the next chapter. Take synaptic weights, $w_1 = 0.8$, connecting A and B to O1; $w_2 = 0.45$, connecting A and B to O2, and $x1 = -1.5$, representing the inhibitory connection from O2 to O1. Using equations (16.4), take $\theta = \gamma = \epsilon = 0.1$, and $m = -100$, and $c = 60$.

Solution. By coupling together four FN equations of the form (16.4), representing inputs A, B, and outputs S and C, the binary oscillator half-adder can be simulated as shown in Program_16b.py. The weights of the excitatory and inhibitory connections are indicated in the program and explained in the PLoS ONE paper. The outputs are shown in Figure 16.4(d).

```python
# Program_16b.py: The Fitzhugh-Nagumo Half-Adder.
import numpy as np
import matplotlib.pyplot as plt
from scipy.integrate import odeint
# Input current
def input_1(t): return 1 * (t > 500) - 1 * (t>1000) + 1 * (t > 1500)
def input_2(t): return 1 * (t > 1000)
# Constants
theta = gamma = epsilon = 0.1
tmax, m, c = 2000, -100, 60
t = np.arange(0.0, 2000.0, 0.1)
# Four coupled Fitzhugh-Nagumo equations.
def fn_odes(X, t):
    u1, v1, u2, v2, u3, v3, u4, v4 = X
    du1 = -u1 * (u1 - theta) * (u1 - 1) - v1 + input_1(t)
    dv1 = epsilon * (u1 - gamma * v1)
    du2 = -u2 * (u2 - theta) * (u2 - 1) - v2 + input_2(t)
    dv2 = epsilon * (u2 - gamma * v2)
    du3 = -u3 * ((u3 - theta) * (u3 - 1) - v3 +
                0.8 / (1 + np.exp(m*v1 + c))   # Excitation.
                + 0.8 / (1 + np.exp(m*v2 + c))   # Excitation.
                - 1.5 / (1 + np.exp(m*v4 + c)))   # Inhibition
    dv3 = epsilon * (u3 - gamma*v3)
    du4 = (-u4 * (u4 - theta) * (u4 - 1) - v4
            + 0.45 / (1 + np.exp(m*v1 + c))   # Excitation.
            + 0.45 / (1 + np.exp(m*v2 + c)))   # Excitation.
    dv4 = epsilon * (u4 - gamma * v4)
    return (du1, dv1, du2, dv2, du3, dv3, du4, dv4)
y0 = [0.01, 0.01, 0.01, 0.01, 0, 0, 0, 0]
X = odeint(fn_odes, y0, t, rtol=1e-6)
u1, v1, u2, v2, u3, v3, u4, v4 = X.T   # Unpack columns
plt.subplots_adjust(hspace=1)
plt.figure(1)
plt.subplot(4, 1, 1)
plt.title("Fitzhugh-Nagumo Half-Adder")
plt.plot(t, u1, "b")
plt.ylim(-1, 1.5)
plt.ylabel("I$_1$")
plt.subplot(4, 1, 2)
```

```
plt.plot(t, u2, "b")
plt.ylim(-1, 1.5)
plt.ylabel("I$_2$")
plt.subplot(4, 1, 3)
plt.plot(t, u3, "g")
plt.ylim(0, 1)
plt.ylim(-1, 1.5)
plt.ylabel("O$_1$")
plt.subplot(4, 1, 4)
plt.plot(t, u4, "g")
plt.ylim(-1, 1.5)
plt.ylabel("O$_2$")
plt.xlabel("Time")
plt.show()
```

16.3 THE BINARY OSCILLATOR SET RESET FLIP-FLOP

For an introduction to hands-on electronics with analog and digital circuits, the reader is directed to the excellent book of Kaplan and White [14], where transistor-based logic and set reset (SR) flip-flop devices can be made on breadboards. In electronics, an SR flip-flop (or latch) is a bistable device that has two stable states and can be used to store memory. There are many different types of flip-flop and they have a wide range of applications as/for bounce elimination switches, counters, data storage, data transfer, frequency division, random access memory (RAM) and shift and storage registers, for example.

A simple SR flip-flop can be constructed using either two NOR or two NAND logic gates. Figure 16.5(a) shows the configuration of an SR flip-flop made of a pair of NOR-logic gates. To cause a switch, the whole line has to be charged, which uses power. Alternatively, an SR flip-flop can be constructed using threshold oscillators, as indicated in Figure 16.5(b). There are many advantages in using the threshold oscillator device over the transistor-based device which will now be highlighted.

Example 16.3.1. Using the FN equations (16.4), simulate the SR flip-flop displayed in Figure 16.5(b).

Solution. By coupling together four FN equations of the form (16.4) and using Program_16c.py, the binary oscillator SR flip-flop is shown to behave correctly, see Figure 16.6. In Figure 16.6(a), inputs I_1 and I_2 are from threshold oscillators, and the device is shown to be functioning correctly. In [2], the authors show that the device also works under ballistic propagation, where a single pulse from a threshold oscillator is all that is required to cause a switch. In this way, power is saved and less power is required to run these devices when compared to transistor-based SR flip-flops. In Figure 16.6(b), inputs I_1 and I_2 are single pulses of electricity, demonstrating that the devices could be connected to CMOS circuits.

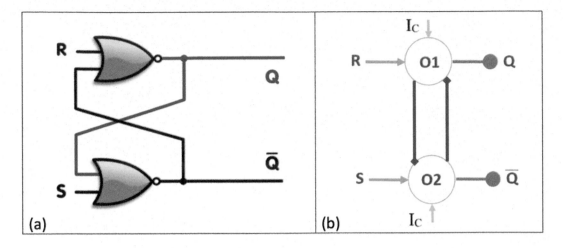

Figure 16.5 Bistable Set Reset Flip-Flops. (a) An SR flip-flop made from a couple of cross-coupled NOR gates. Symbols $R, S, Q, \overline{Q}$ stand for reset, set, current output Q, and not Q, respectively. If Q is on, $\overline{Q}$ is off, and vice-versa. (b) An SR flip-flop constructed from two threshold oscillators O1 and O2. The current I_C keeps the oscillators close to threshold.

```python
# Program_16c.py: Fitzhugh-Nagumo Set Reset Flip-Flop.
import numpy as np
import matplotlib.pyplot as plt
from scipy.integrate import odeint
# Input currents.
Ic = 1
```

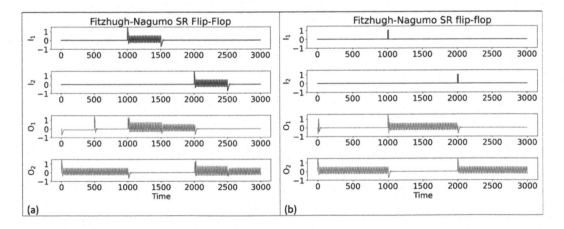

Figure 16.6 Output from binary oscillator SR flip-flops. In both cases, $Ic = 1$, and the devices are performing in the correct manner. (a) The output from four coupled FN oscillators. (b) The output from two coupled FN oscillators, O1 and O2, where I_1 and I_2 are simple current pulses.

```python
def Input_1(t):
    return 1*(t>1000)-1*(t>1500)
def Input_2(t):
    return 1*(t>2000)-1*(t>2500)
def Ic1(t):
    return Ic*(t>10)
def Ic2(t):
    return Ic*(t>0)
# Constants
theta,gamma,epsilon,m,c,Tmax = 0.1,0.1,0.1,-100,60,2000
w1 , x1 = 0.5 , -1
t=np.arange(0.0, 3000.0, 0.1)
# Inputs from neurons and outputs for O1 and O2.
def FN_ODEs(X,t):
    u1,v1,u2,v2,u3,v3,u4,v4=X
    du1 = -u1*(u1-theta)*(u1-1)-v1+Input_1(t)
    dv1 = epsilon*(u1-gamma*v1)
    du2 = -u2*(u2-theta)*(u2-1)-v2+Input_2(t)
    dv2 = epsilon*(u2-gamma*v2)
    du3 = -u3*(u3-theta)*(u3-1)-v3+w1/(1+np.exp(m*v1+c)) \
              +x1/(1+np.exp(m*v4+c)) + Ic1(t)
    dv3 = epsilon*(u3-gamma*v3)
    du4 = -u4*(u4-theta)*(u4-1)-v4+0.45/(1+np.exp(m*v2+c)) \
              +x1/(1+np.exp(m*v3+c)) + Ic2(t)
    dv4 = epsilon*(u4-gamma*v4)
    return du1,dv1,du2,dv2,du3,dv3,du4,dv4

X = odeint(FN_ODEs,[0.01,0.01,0.01,0.01,0,0,0,0],t,rtol=1e-6)
u1,v1,u2,v2 = X[:,0],X[:,1],X[:,2],X[:,3]
u3,v3,u4,v4 = X[:,4],X[:,5],X[:,6],X[:,7]
plt.rcParams["font.size"] = "20"
plt.subplots_adjust(hspace = 1)
plt.figure(1)
plt.subplot(4,1,1)
plt.title("Fitzhugh-Nagumo SR Flip-Flop")
plt.plot(t, u1, "b")
plt.ylim(-1, 1.5)
plt.ylabel("I$_1$")
plt.subplot(4,1,2)
plt.plot(t, u2, "b")
plt.ylim(-1, 1.5)
plt.ylabel("I$_2$")
plt.subplot(4,1,3)
plt.plot(t, u3, "g")
plt.ylim(0, 1)
```

```
plt.ylim(-1, 1.5)
plt.ylabel("O$_1$")
plt.subplot(4,1,4)
plt.plot(t, u4, "g")
plt.ylim(-1, 1.5)
plt.ylabel("O$_2$")
plt.xlabel("Time")
plt.show()
```

16.4 REAL-WORLD APPLICATIONS AND FUTURE WORK

A typical human brain consists of approximately 10^{11} neurons connected by about 10^{15} synapses. It has been estimated that the average brain consumes just 25 Watts of power and is extremely energy efficient. IBM have estimated that if a transistor-based computer could be built to mimic one human brain, then it would require hundreds of megaWatts of power, thus demonstrating the inefficiencies of CMOS technology. Using oscillators to perform computation is not a new idea, indeed, in 1948, the first programmable computer in the world, the Baby, was made in Manchester, UK, and one of the principal components used in its construction was the vacuum tube oscillator. The Museum of Science and Industry (MOSI), based in Manchester, currently houses a full working replica of the Baby computer—I recommend a visit. With the advent of the transistor in the 1950s, the power inefficient vacuum tubes quickly became obsolete in computing, and modern day computers are based on switching technology using complementary metal-oxide-semiconductor (CMOS) technology. CMOS has served the world well for many decades, however, scaling bulk CMOS below 100 nanometres is causing manufacturing problems and new technologies are being sought for the next generation of computers [3]. Using threshold oscillator devices in computing would be a paradigm shift, and the author and his collaborators are working toward this goal.

Using threshold oscillators, the processing power can be doubled with a linear increase in components, see [2] and the Exercises at the end of the chapter. This is a major advantage over CMOS, where a doubling of processing power requires an exponential increase in the number of transistors required. That is why current supercomputers are composed of thousands of trillions of transistors which consume vast amounts of energy. As far as memory is concerned, low-power ballistic propagation-based data interconnect between processor and memory is feasible, which again saves massively on power consumption.

There are potentially five avenues of research for binary oscillator computing, including:

- **Josephson Junction (JJ) Oscillators.** JJs are superconducting natural threshold oscillators that cycle one hundred million times faster and can be manufactured one thousand times smaller than biological neurons.

- **Biological Neuron Oscillators.** The proposal is not to build a new type of computer using these devices, however, simple logic and memory circuits could be used to model neuronal degradation.

- **CMOS Oscillators.** In 2017, Sourikopoulos et al. [22] designed an artificial neuron in standard 65 nanometre CMOS technology, where they emphasize the ultra-low energy consumption and discuss applications in building artificial neural networks. To-date, little research has been conducted on binary oscillator computing using CMOS and the author is seeking collaborators for this work.

- **Memristors.** Neuristors act like neurons and are manufactured using memristors [20], and recently, memristor synapses and neurons with structural plasticity have been investigated [23]. In a recent paper, Fang et al. [8] have shown how memristors can be used to perform binary threshold oscillator logic. They simulate a Fitzhugh-Nagumo model using a memristor instead of a nonlinear resistor and the synaptic connectors are memristors.

- **Optical Oscillators.** Photonic synapses are employed for brain-like computing [11].

The author and his collaborators have made progress with JJs and biological neurons and new research papers will be published in the coming years.

Binary Oscillator Computing with JJs. In 2010, Ken Segall and his collaborators published a paper with the enthralling title, JJ simulation of neurons [6], and in 2017, his group built physical devices that demonstrated synchronization dynamics on the picosecond timescale for coupled JJ neurons [21]. We have been working with Ken for a number of years and have recently obtained some exciting results on JJ set reset flip-flops. The circuitry for a JJ SR flip-flop is shown in Figure 16.7. We have shown that the devices run incredibly quickly and that oscillating memory is more stable than its partner static memory. For a stability analysis of a JJ neuron, the reader is directed to [5].

Building an Assay for Neuronal Degradation. There are known to be over five hundred neurological disorders that can affect the human body including, headaches, epilepsy and seizures, stroke, Alzheimer's disease and dementia, and Parkinson's disease, for example. The author and his collaborators at Loughborough University, Aston University, and University of Cambridge, are proposing to build assays for neuronal degradation, where simple neural architectures are trained to act as half-adders, full-adders, and SR flip-flop memory circuits. Stem cells can then be used to grow excitatory and inhibitory neurons, which can be connected together using chemical techniques. Diseased neurons can be used, and the functionality of the simple circuits can be measured when subjected to certain drugs. This research is in its very early stages. Figures 16.8(a) and (b) show the MEA-Lite apparatus and single neurons sat atop a multi-electrode-array (MEA)—connected by axons, respectively. Mathematical models of these and other devices are covered in [17] and more detail is provided there. We are working toward making viable devices for medicinal applications in the near future.

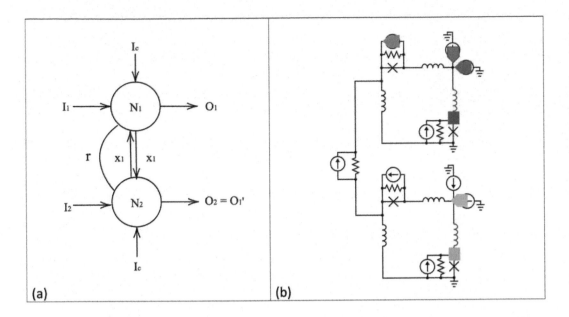

Figure 16.7 (a) Schematic of an SR flip-flop. (b) The JJ circuitry for an SR flip-flop. The X symbols represent JJs.

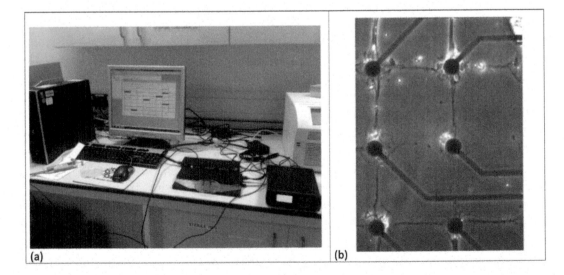

Figure 16.8 (a) MEA-Lite equipment. The computer screen is showing that some of the neurons are firing. (b) A magnification of individual neurons sat atop an MEA. Used with the permission of C. Wyart.

EXERCISES

16.1 Consider the Fitzhugh-Nagumo model of a neuron given by equation (16.4). Take $\theta = 0.14$, $\epsilon = 0.01$ and $\gamma = 2.54$. Produce an animation of a phase portrait in the (u, v) plane as the input current i increases from $i = 0$ to $i = 0.2$. Choose $u_0 = 0$ and $v_0 = 0.1$ You should see a limit bifurcation from a critical point which grows and then shrinks back to a critical point.

16.2 Consider the dimensionless Morris-Lecar model of a neuron [18] given by the ODEs:

$$C\frac{dV}{dt} = i - g_{fast}M_{SS}(V)(V - E_{Na}) - g_{slow}N(V - E_K) - g_{leak}(V - E_{leak})$$
$$\frac{dN}{dt} = \phi\frac{N_{SS}(V) - N}{\tau_N(V)},$$

where

$$M_{SS}(V) = 0.5\left(1 + \tanh\left(\frac{V - V_1}{V_2}\right)\right),$$
$$N_{SS}(V) = 0.5\left(1 + \tanh\left(\frac{V - V_3}{V_4}\right)\right),$$
$$\frac{1}{\tau_N(V)} = \cosh\left(\frac{V - V_3}{2V_4}\right),$$

where V is membrane potential, N is a recovery variable, i is applied current, C is membrane capacitance, $g_{fast}, g_{slow}, g_{leak}$, represent conductance through membrane channels, E_K, E_{Na}, E_{leak}, are equilibrium potentials of the relevant channels, ϕ, V_1, V_2, V_3, V_4 are constants. Use **odeint** to show that there are three limit cycles in the (V, N) plane when, $i = 82$, $g_{fast} = 20$, $V_1 = -1.2$, $V_2 = 18$, $V_3 = -20.5$, $V_4 = 10$, $E_{Na} = 50$, $E_K = -100$, $g_{slow} = 20$, $g_{leak} = 2$, $E_{leak} = -70$, $\phi = 0.15$ and $C = 2$.

16.3 Reproduce the results shown in Figure 16.9(b) for the binary oscillator full-adder, see Figure 16.9(a), which is used to add three bits together.

16.4 Demonstrate that the SR flip-flop shown in Figure 16.5(b) works under ballistic propagation, where a single pulse from a neuron is sufficient to cause a switch.

Solutions to the Exercises may be found here:

https://drstephenlynch.github.io/webpages/Solutions_Section_3.html.

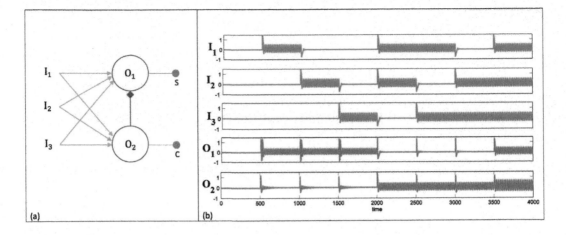

Figure 16.9 (a) Schematic of a binary oscillator full-adder. (b) Correct time series for a full-adder. In the final column $1 + 1 + 1 = 11$, in binary.

FURTHER READING

[1] Berner, R. (2021). *Patterns of Synchrony in Complex Networks of Adaptively Coupled Oscillators (Springer Theses)*. Springer, New York.

[2] Borresen, J. and Lynch, S. (2012). Oscillatory Threshold Logic. *PLoS ONE*. **7**(11): e48498.

[3] Borresen, J. and Lynch, S. (2009). Neuronal computers. *Nonlinear Anal. Theory, Meth. and Appl.* **71**, 2372–2376.

[4] Bucaro, S. (2019). *Basic Digital Logic Design: Use Boolean Algebra, Karnaugh Mapping, or an Easy Free Open-Source Logic Gate Simulator*. bucarotechelp.com.

[5] Chalkiadakis, D. and Hizanidis, J. (2022) Dynamical properties of neuromorphic Josephson junctions. *Phys. Rev. E* **106**, 044206.

[6] Crotty, P., Schult, D. and Segall, K. (2010). Josephson junction simulation of neurons. *Phys. Rev. E* **82** 011914.

[7] Destexhe, A., Mainen, Z.F. and Sejnowski, T.J. (1994). An efficient method for computing synaptic conductances based on a kinetic model of receptor binding. *Neural Computation*. **6**, 14–18.

[8] Fang, X., Duan, S. and Wang, L. (2022). Memristive FHN spiking neuron model and brain-inspired threshold logic computing. *Neurocomputing*. **517**, 93–105.

[9] Feynman, R.P. (2000). *Feynman Lectures on Computation*. CRC Press, Boca Raton, FL.

[10] Fitzhugh, R. (1961). Impulses and physiological states in theoretical models of nerve membranes, *Biophys*. **1182**, 445–466.

[11] Gholipour, B., Bastock, P., Craig, C. et al. (2015). Amorphous metal-sulphide microfibres enable photonic synapses for brain-like computing. *Advanced Optical Materials*. **3** 635–641.

[12] Harris, S.L. and Harris, D. (2021). *Digital Design and Computer Architecture, RISC-V Edition*. Morgan Kauffmann, San Francisco, CA.

[13] Hodgkin, A.L. and Huxley, A.F. (1952). A qualitative description of membrane current and its application to conduction and excitation in nerve. *J. Physiol.* **117**, 500-544. Reproduced in *Bull. Math. Biol.* **52**, (1990), 25–71.

[14] Kaplan, D.M. and White, C.G. (2003). *Hands-On Electronics: A Practical Introduction to Analog and Digital Circuits*. Cambridge University Press, Cambridge, UK.

[15] Lynch, S. and Borresen, J. (2012). Binary half-adder using oscillators. *International Publication. Number, WO 2012/001372 A1*. 1–57.

[16] Lynch, S., Borresen, J. and Slevin, M.A. (2015). International patent: Cell assays, development of binary half-adder. *Publication Number US 2015/0276707 A1*.

[17] Lynch, S., Borresen, J., Roach, P. et al. (2020). Mathematical modeling of neuronal logic, memory and clocking circuits. *Int. J. of Bifurcation and Chaos.* **30** 2050003, 1–16.

[18] Morris, C. and Lecar, H. (1981). Voltage oscillations in the barnacle giant muscle fiber. *Biophys. J.* **35**(1) 193–213.

[19] Nagumo, J., Arimoto, S. and Yoshizawa, S. (1970). An active pulse transmission line simulating 1214-nerve axons, *Proc. IRL.* **50**, 2061–2070.

[20] Picket, M.D., Medeiros-Ribeiro, G. and Williams, R.S. (2013). A scalable neuristor built with Mott memristors. *Nature Materials.* **12** 114.

[21] Segall, K., LeGro, M., Kaplan, S. et al. (2017).Synchronization dynamics on the picosecond timescale in coupled Josephson junction neurons. *Phys. Rev. E* **95** 032220.

[22] Sourikopoulos, I., Hedayat, S., Loyez, C. et al. (2017). A 4-fJ/spike artificial neuron in 65 nm CMOS technology. *Front. Neurosci.* **11** 123. doi:10.3389/fnins.2017.00123.

[23] Su, B., Cai, J., Wang, Z. et al. (2022). A π-type memristor synapse and neuron with structural plasticity. *Front. Phys.* 10.3389/fphy.2021.798971.

## CHAPTER 17

Neural Networks and Neurodynamics

Neural networks, or artificial neural networks (ANNs), are at the heart of machine learning, deep learning, and artificial intelligence. They are inspired by human brain dynamics, as discussed in the previous chapter, and mimic the way that biological neurons communicate with one another. The first section introduces a brief history of ANNs and the basic theory is applied to simple logic gates. The second section covers feedforward and backpropagation, and the third section demonstrates how to train a simple neural network to value houses. The final section provides an introduction to neurodynamics using simple discrete models of neuromodules. A stability diagram is plotted which shows parameter regions where a system is in steady-state (period one), and unstable, bistable, and quasiperiodic (almost periodic) states.

For a more in-depth introduction to neural networks, the reader is directed to the excellent texts [1], [7], [10], and for an introduction with Python, see [2] and [4].

Aims and Objectives:

- To provide a brief historical introduction.

- To investigate simple architectures.

- To introduce neurodynamics.

At the end of the chapter, the reader should be able to

- construct simple neural networks for logic gates;

- use feedforward and backpropagation to train networks;

- study the dynamics of neuromodules in terms of bistability, chaos, periodicity and quasiperiodicity.

17.1 HISTORY AND THEORY OF NEURAL NETWORKS

Alan Turing is widely acknowledged as the father of theoretical computer science and artificial intelligence. In 1937, he published a famous paper on computable numbers

DOI: 10.1201/9781003285816-17

with an application to the decision problem [11]. In 1950, he published the seminal paper on computing machinery and intelligence [12]. The all-or-none law model of a neuron devised by McCulloch and Pitts [5] in 1943, is the seminal paper on neural networks. They showed, in principle, that a neuron could compute any arithmetic or logic function, and even today the McCulloch-Pitts model of a neuron is the one most widely used. In 1949, Hebb [3] proposed the first learning law for neural networks and there have been many variations to-date. In 1958, Rosenblatt [8] first introduced the perceptron and the perceptron learning rule. Certain problems were highlighted by Minsky and Papert [6] in 1969, and these problems were mainly addressed by Rumelhart, Hinton and Williams [8] in 1986, by introducing the so-called backpropagation algorithm from mathematics. Since 1986, there have been huge advances in the field, which will be covered in more detail in later chapters of the book. Figure 17.1 shows a schematic of a mathematical (artificial) neuron.

The mathematical model of the neuron shown in Figure 17.1 is described by the following equations:

$$v = \sum_{i=1}^{n} x_i w_i + b, \quad y = \sigma(v), \tag{17.1}$$

where v is an activation potential, x_i are inputs, w_i are weights, b is a bias, σ is a transfer function, and y gives the output of the neuron.

In this chapter, the sigmoid (logistic) and tanh transfer functions are used. Other transfer functions will be introduced later in the book. The sigmoid transfer function is defined by:

$$\sigma(v) = \frac{1}{1 + e^{-v}}, \tag{17.2}$$

where v is an activation potential. The sigmoid function is unipolar as it maps to the interval $(0, 1)$. For the sigmoid function, it is not difficult to show that,

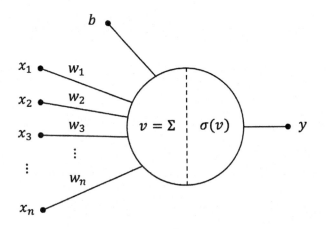

Figure 17.1 Schematic of a mathematical model of a neuron. Notice the similarity to Figure 16.1. The inputs x_i, and the bias b, are connected to the cell body via dendrites of synaptic weights w_i, v is the activation potential of the neuron, σ is a transfer function, and y is the output.

$\frac{d\sigma(v)}{dv} = \sigma(v)(1 - \sigma(v))$, which is useful when programming for backpropagation, as the reader will see in the next section. The tanh transfer function is defined by:

$$\phi(v) = \tanh(v) = \frac{e^v - e^{-v}}{e^v + e^{-v}}, \tag{17.3}$$

where v is an activation potential. The tanh function is bipolar as it maps to the interval $(-1, 1)$. For the tanh function, it is not difficult to show that, $\frac{d\phi(v)}{dv} = 1 - (\phi(v))^2$.

Simple neural networks of AND, OR, and XOR logic gates will now be investigated.

Example 17.1.1. Figure 17.2(a) shows a simple ANN for the logic AND gate, and Figure 17.2(b) lists the corresponding truth table. Write a Python program to illustrate that the ANN does indeed act as a good approximation of an AND gate using the sigmoid transfer function.

Solution. Using equations (17.1) and (17.2), Program_17a.py shows that the ANN in Figure 17.2(a) gives a very good approximation of an AND gate. The output is AND(0,0)=9.357622968839299e-14, AND(1,0)=4.5397868702434395e-05, AND(0,1)=4.5397868702434395e-05, AND(1,1)=0.9999546021312976.

```
# Program_17a.py: ANN for an AND Gate.
import numpy as np
w1 , w2 , b = 20 , 20 , -30
def sigmoid(v):
    return 1 / (1 + np.exp(- v))
def AND(x1, x2):
    return sigmoid(x1 * w1 + x2 * w2 + b)
print("AND(0,0)=", AND(0,0))
print("AND(1,0)=", AND(1,0))
print("AND(0,1)=", AND(0,1))
print("AND(1,1)=", AND(1,1))
```

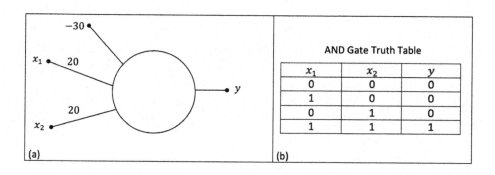

Figure 17.2 (a) ANN for an AND gate. (b) Truth table for an AND gate.

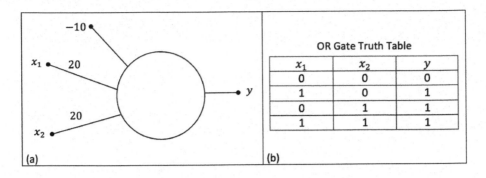

Figure 17.3 (a) ANN for an OR gate. (b) Truth table for an OR gate.

Example 17.1.2. Figure 17.3(a) shows a simple ANN for the logic OR gate, and Figure 17.3(b) lists the corresponding truth table. Write a Python program to illustrate that the ANN does indeed act as a good approximation of an OR gate using the sigmoid transfer function.

Solution. Program_17b.py shows that the ANN in Figure 17.3(a) gives a very good approximation of an OR gate. The output is OR(0,0)=4.5397868702434395e-05, OR(1,0)=0.9999546021312976, OR(0,1)=0.9999546021312976, OR(1,1)=0.9999999999999065.

```
# Program_17b.py: ANN for an OR Gate.
import numpy as np
w1 , w2 , b = 20 , 20 , -10
def sigmoid(v):
    return 1 / (1 + np.exp(- v))
def OR(x1, x2):
    return sigmoid(x1 * w1 + x2 * w2 + b)
print("OR(0,0)=", OR(0,0))
print("OR(1,0)=", OR(1,0))
print("OR(0,1)=", OR(0,1))
print("OR(1,1)=", OR(1,1))
```

The solutions for the AND and OR gates are linearly separable and only one neuron is required to solve these problems, however, the solutions for the exclusive OR (XOR) gate are not linearly separable, and a hidden layer must be introduced. Figure 17.4(a) shows the ANN and Figure 17.4(b) displays the truth table for an XOR logic gate. Given that, $w_{11} = 60$, $w_{12} = 80$, $w_{21} = 60$, $w_{22} = 80$, $w_{13} = -60$, $w_{23} = 60$, $b_1 = -90$, $b_2 = -40$ and $b_3 = -30$, and using the sigmoid transfer function, the ANN acts as an XOR logic gate. The reader is asked to verify this in the Exercises at the end of the chapter.

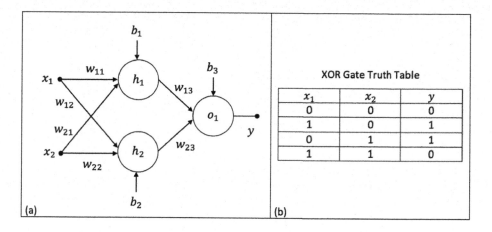

Figure 17.4 (a) ANN for an XOR gate, the synaptic weights are given in the text. There is one hidden layer consisting of two neurons, labelled h_1 and h_2, and there is one output neuron, o_1. Here, h_1, h_2, o_1, are activation potentials. (b) The truth table for an XOR gate. The problem is not linearly separable when only one neuron is used.

17.2 THE BACKPROPAGATION ALGORITHM

In the examples considered thus far, the weights for the ANNs have been given, but what if we are not given those weights? The backpropagation of errors algorithm is used to determine those weights. To illustrate the method, consider Figure 17.5 and use the transfer function $\sigma(v)$, furthermore suppose that the biases b_1 and b_2 are constant. Suppose we wish to update weights w_2 and w_1.

Initially, random weights are chosen, data is fed forward through the network, and an error inevitably results. In this example, we use a mean squared error. Next, the backpropagation algorithm is used to update the weights in order to minimize the error. Thus, using the chain rule,

$$\frac{\partial Err}{\partial w_2} = \frac{\partial Err}{\partial y}\frac{\partial y}{\partial w_2} = \frac{\partial Err}{\partial y}\frac{\partial y}{\partial o_1}\frac{\partial o_1}{\partial w_2},$$

and

$$\frac{\partial Err}{\partial w_2} = (y_t - y) \times \frac{\partial \sigma}{\partial o_1} \times \sigma(h_1),$$

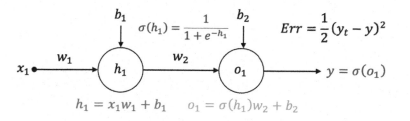

Figure 17.5 Simple ANN to illustrate the backpropagation algorithm, where Err represents the error and y_t is the target output.

and

$$\frac{\partial Err}{\partial w_2} = (y_t - y) \times \sigma(o_1)(1 - \sigma(o_1)) \times \sigma(h_1).$$

The weight w_2 is then updated using the formula:

$$w_2 = w_2 - \eta \times \frac{\partial Err}{\partial w_2},$$

where η is called the learning rate.

A similar argument is used to update w_1, thus:

$$\frac{\partial Err}{\partial w_1} = \frac{\partial Err}{\partial y}\frac{\partial y}{\partial w_1} = \frac{\partial Err}{\partial y} \times \frac{\partial y}{\partial o_1} \times \frac{\partial o_1}{\partial \sigma(h_1)} \times \frac{\partial \sigma(h_1)}{\partial h_1} \times \frac{\partial h_1}{\partial w_1},$$

and

$$\frac{\partial Err}{\partial w_1} = (y_t - y) \times \sigma(o_1)(1 - \sigma(o_1)) \times w_2 \times \sigma(h_1)(1 - \sigma(h_1)) \times x_1.$$

The weight w_1 is then updated using the formula:

$$w_1 = w_1 - \eta \times \frac{\partial Err}{\partial w_1}.$$

Example 17.2.1. Consider the ANN displayed in Figure 17.4(a). Given that, $w_{11} = 0.2$, $w_{12} = 0.15$, $w_{21} = 0.25$, $w_{22} = 0.3$, $w_{13} = 0.15$, $w_{23} = 0.1$, $b_1 = b_2 = b_3 = -1$, $x_1 = 1$, $x_2 = 1$, $y_t = 0$ and $\eta = 0.1$:

(i) Determine the output y of the ANN on a forward pass given that the transfer function is $\sigma(x)$.

(ii) Use the backpropagation algorithm to update the weights w_{13} and w_{23}.

Solution. Program_17c.py gives $y = 0.28729994077761756$, and new weights $w_{13} = 0.1521522763401024$, and $w_{23} = 0.10215420563664182$. The reader will be asked to update the other weights in the Exercises.

```
# Program_17c.py: Simple Backpropagation.
import numpy as np
w11,w12,w21,w22,w13,w23 = 0.2,0.15,0.25,0.3,0.15,0.1
b1 , b2 , b3 = -1 , -1 , -1
yt , eta = 0 , 0.1
x1 , x2 = 1 , 1
def sigmoid(v):
    return 1 / (1 + np.exp(- v))
h1 = x1 * w11 + x2 * w21 + b1
h2 = x1 * w12 + x2 * w22 + b2
o1 = sigmoid(h1) * w13 + sigmoid(h2) * w23 + b3
y = sigmoid(o1)
```

```
print("y = ", y)
# Backpropagate.
dErrdw13=(yt-y)*sigmoid(o1)*(1-sigmoid(o1))*sigmoid(h1)
w13 = w13 - eta * dErrdw13
print("w13 = ", w13)
dErrdw23=(yt-y)*sigmoid(o1)*(1-sigmoid(o1))*sigmoid(h2)
w23 = w23 - eta * dErrdw23
print("w23 = ", w23)
```

17.3 MACHINE LEARNING ON BOSTON HOUSING DATA

Machine Learning (ML) is a subset of artificial intelligence that enables systems to learn and improve from experience without human intervention. As a simple example, consider the famous problem of valuing houses in Boston's suburbs. The perceptron shown in Figure 17.1 is all that is required to solve this problem. The famous data set was collected in 1978 and lists 506 entries (houses) which each have 14 attributes, from various localities in the Boston area. Included in the attributes are accessibility to radial highways, per capita crime rate by town, nitric oxide concentrations, pupil-teacher ratio by town, percentage lower status of the population and weighted distances to five Boston employment centres, for example. At this stage it is worth taking a few minutes to ponder on what attributes would be important to you when purchasing a house. Different attributes will have higher priorities for different people, however, overall the value of houses can be predicted quite accurately if enough data is provided. Surprisingly, rather little data is required to accurately value homes. The data can be downloaded from the UCI Machine Learning Repository, or from the book's GitHub site, where the file is called housing.txt. The aim is to minimize an error function with respect to the weights of the synaptic connections in the ANN. In most cases, the mean squared error is used, as in Example 17.2.1. This is an unconstrained optimization problem, where the weights, w_k say, are changed to minimize the error. The method of steepest descent is used and the weights are updated using the so-called generalized delta rule. There are two ways to update the weights, they can be updated instantaneously on each iteration (asynchronous), or they can be updated in batches, where weights are updated at the same time, based on the average of one so-called epoch (synchronous). An epoch is the number of passes of one complete data set, thus, for the Boston housing data set an epoch entails 506 iterations. By feeding through the entire dataset a number of times, the ANN can be trained and the weights should converge to set values. The number of epochs needed and the learning rate must be determined from experimentation. Once the ANN is trained, it can be used to give accurate values of houses from the dataset, and can be improved and updated by adding more datasets. To aid understanding of the program, the backpropagation algorithm can be broken down into five steps:

1. Scale the data to zero mean, unit variance and introduce a bias on the input.

2. Set small random weights.

3. Set the number of epochs and learning rate.

4. Compute outputs y_k, the errors $t_k - y_k$, the gradients $g_k = \frac{\partial Err}{\partial w_k}$, and perform gradient descent to evaluate, $w_k(n+1) = w_k(n) - \eta g_k$, for each weight.

5. Plot a graph of weights versus number of iterations.

This is an example of supervised learning where the input and target output values are known. The majority of machine learning applications use supervised learning. Unsupervised learning on the other hand is used when there is no target data.

Example 17.3.1. Write a Python program to create an ANN for the Boston Housing data for three attributes: column six (average number of rooms), nine (index of accessibility to radial highways), and 13 (percentage lower status of population), using the target data presented in column 14 (median value of owner-occupied homes in thousands of dollars). Use the activation function $\phi(v)$, asynchronous updating and show how the weights are adjusted as the number of iterations increases.

Solution. By experimentation, it was found that the weights converged after 100 epochs, 50600 iterations, and a learning rate of $\eta = 0.0005$, was sufficient for this problem. Program_17d.py gives Figure 17.6, showing how the weights converge.

```
# Program_17d.py: Boston Housing Data.
import matplotlib.pyplot as plt
import numpy as np
data = np.loadtxt("housing.txt")
rows, columns = data.shape
columns = 4  # Using 4 columns from the dataset in this case.
X , t = data[:, [5, 8, 12]] , data[:, 13]
ws1, ws2, ws3, ws4 = [], [], [], []
# Normalize the data.
xmean , xstd = X.mean(axis=0) , X.std(axis=0)
ones = np.array([np.ones(rows)])
X = (X - xmean * ones.T) / (xstd * ones.T)
X = np.c_[np.ones(rows), X]
tmean , tstd = (max(t) + min(t)) / 2 , (max(t) - min(t)) / 2
t = (t - tmean) / tstd
# Set random weights.
w = 0.1 * np.random.random(columns)
y1 = np.tanh(X.dot(w))
e1 = t - y1
mse = np.var(e1)
num_epochs , eta = 100 , 0.0005
k = 1
for m in range(num_epochs):
```

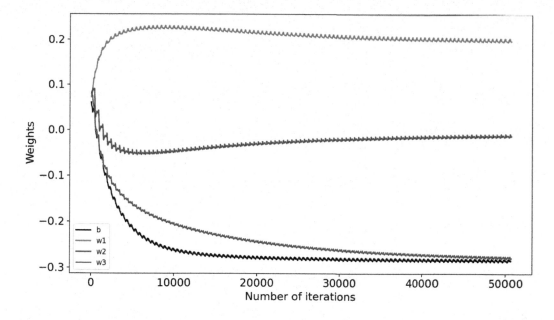

Figure 17.6 See Figure 17.1 for the ANN. Graph of weights versus number of iterations. In this case, 100 epochs were used (50600 iterations) and the learning rate was set at $\eta = 0.0005$. There is clear convergence.

```
for n in range(rows):
    yk = np.tanh(X[n, :].dot(w))
    err = yk - t[n]
    g = X[n, :].T * ((1 - yk**2) * err) # Gradient vector.
    w = w - eta*g                       # Update weights.
    k += 1
    ws1.append([k, np.array(w[0]).tolist()])
    ws2.append([k, np.array(w[1]).tolist()])
    ws3.append([k, np.array(w[2]).tolist()])
    ws4.append([k, np.array(w[3]).tolist()])
ws1,ws2,ws3,ws4=np.array(ws1),np.array(ws2),np.array(ws3),np.array(ws4)
plt.plot(ws1[:, 0],ws1[:, 1],"k",markersize=0.1,label="b")
plt.plot(ws2[:, 0],ws2[:, 1],"g",markersize=0.1,label="w1")
plt.plot(ws3[:, 0],ws3[:, 1],"b",markersize=0.1,label="w2")
plt.plot(ws4[:, 0],ws4[:, 1],"r",markersize=0.1,label="w3")
plt.xlabel("Number of iterations", fontsize=15)
plt.ylabel("Weights", fontsize=15)
plt.tick_params(labelsize=15)
plt.legend()
plt.show()
```

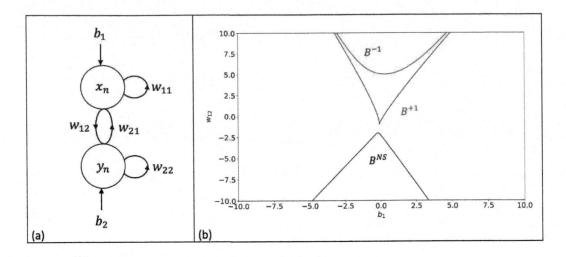

Figure 17.7 (a) Two-neuron module. (b) Stability diagram in the (b_1, w_{12}) plane when $b_2 = -1$, $w_{11} = 1.5$, $w_{21} = 5$, $\alpha = 1$ and $\beta = 0.1$. B^{+1} is the bistable boundary curve, where the system displays hysteresis, B^{-1} is the unstable boundary curve, where the system is not in steady state, and B^{NS} is the Neimark-Sacker boundary curve, where the system can show quasiperiodic behavior.

17.4 NEURODYNAMICS

This section introduces the reader to neurodynamics and determining stability regions for dynamical systems by means of a simple example. Figures 17.7(a) and 17.7(b) show a two-neuron module and corresponding stability diagram, respectively. The discrete dynamical system that models the two-neuron module is given as:

$$x_{n+1} = b_1 + w_{11}\phi_1(x_n) + w_{12}\phi_2(y_n), \quad y_{n+1} = b_2 + w_{21}\phi_1(x_n) + w_{22}\phi_2(y_n), \quad (17.4)$$

where b_1, b_2 are biases, w_{ij} are weights, x_n, y_n are activation potentials, and the transfer functions are $\phi_1(x) = \tanh(\alpha x)$, $\phi_2(y) = \tanh(\beta y)$.

Example 17.4.1. To simplify the analysis, assume that $w_{22} = 0$ in equations (17.4). Take $b_2 = -1$, $w_{11} = 1.5$, $w_{21} = 5$, $\alpha = 1$ and $\beta = 0.1$ Use a stability analysis from dynamical systems theory to determine parameter values where the system is stable, unstable, and quasiperiodic. Use Python to obtain Figure 17.7(b).

Solution. The fixed points of period one, or steady-states, satisfy the equations $x_{n+1} = x_n = x$, say, and $y_{n+1} = y_n = y$, say. Thus,

$$b_1 = x - w_{11}\tanh(\alpha x) - w_{12}\tanh(\beta y), \quad y = b_2 + w_{21}\tanh(\alpha x). \quad (17.5)$$

Use the Jacobian matrix to determine stability conditions. Take $x_{n+1} = P$ and $y_{n+1} = Q$ in equation (17.4), then

$$J = \begin{pmatrix} \frac{\partial P}{\partial x} & \frac{\partial P}{\partial y} \\ \frac{\partial Q}{\partial x} & \frac{\partial Q}{\partial y} \end{pmatrix} = \begin{pmatrix} \alpha w_{11}\text{sech}^2(\alpha x) & \beta w_{12}\text{sech}^2(\beta y) \\ \alpha w_{12}\text{sech}^2(\alpha x) & 0 \end{pmatrix}.$$

The stability conditions of the fixed points are determined from the eigenvalues, and the trace and determinant of the Jacobian matrix. The characteristic equation is given by $\det(J - \lambda I) = 0$, which gives:

$$\lambda^2 - \alpha w_{11}\text{sech}^2(\alpha x)\lambda - \alpha\beta w_{12}w_{21}\text{sech}^2(\alpha x)\text{sech}^2(\beta y) = 0. \qquad (17.6)$$

The fixed points undergo a fold bifurcation (indicating bistability) when $\lambda = +1$. The boundary curve is labelled B^{+1} in Figure 17.7(b). In this case, equation (17.6) gives

$$w_{12} = \frac{1 - \alpha w_{11}\text{sech}^2(\alpha x)}{\alpha\beta w_{12}w_{21}\text{sech}^2(\alpha x)\text{sech}^2(\beta y)}. \qquad (17.7)$$

The fixed points undergo a flip bifurcation (indicating instability) when $\lambda = -1$. The boundary curve is labelled B^{-1} in Figure 17.7(b). In this case, equation (17.6) gives

$$w_{12} = \frac{1 + \alpha w_{11}\text{sech}^2(\alpha x)}{\alpha\beta w_{12}w_{21}\text{sech}^2(\alpha x)\text{sech}^2(\beta y)}. \qquad (17.8)$$

The fixed points undergo a so-called Neimark-Sacker bifurcation (indicating quasiperiodicity), when $\det(J) = 1$, and $|\text{trace}(J)| < 2$. The boundary curve is labelled B^{NS} in Figure 17.7(b). Equation (17.6) gives

$$w_{12} = -\frac{1}{\alpha\beta w_{12}w_{21}\text{sech}^2(\alpha x)\text{sech}^2(\beta y)}. \qquad (17.9)$$

```
# Program_17e.py: Stability Diagram of a Neuromodule.
import numpy as np
import matplotlib.pyplot as plt
# Set parameters.
b2 , w11 , w21 , alpha , beta = -1 , 1.5 , 5 , 1 , 0.1
xmin=5
x=np.linspace(-xmin,xmin,1000)
y=b2 + w21 * np.tanh(x)
def sech(x):
    return 1 / np.cosh(x)
w12=(1-alpha*w11*(sech(alpha*x))**2) / \
    (alpha*beta*w21*(sech(alpha*x))**2*(sech(beta*y))**2)
b1=x-w11*np.tanh(alpha*x)-w12*np.tanh(beta*y)
plt.plot(b1, w12, "b") # Bistable boundary.
w12=(1+alpha*w11*(sech(alpha*x))**2) / \
    (alpha*beta*w21*(sech(alpha*x))**2*(sech(beta*y))**2)
b1=x-w11*np.tanh(alpha*x)-w12*np.tanh(beta*y)
plt.plot(b1, w12, "r") # Unstable boundary.
w12=(-1) / \
    (alpha*beta*w21*(sech(alpha*x))**2*(sech(beta*y))**2)
b1=x-w11*np.tanh(alpha*x)-w12*np.tanh(beta*y)
```

```
plt.plot(b1, w12, "k") # Neimark-Sacker boundary.
plt.rcParams["font.size"] = "20"
plt.xlim(-10,10)
plt.ylim(-10,10)
plt.xlabel("$b_1$")
plt.ylabel("$w_{12}$")
plt.show()
```

Equations (17.5) to (17.9) are required to plot Figure 17.7(b). In order to understand the stability diagram, the reader is encouraged to attempt question 17.4 in the Exercises.

EXERCISES

17.1 Show that the ANN illustrated in Figure 17.4(a) acts as a good approximation of an XOR gate, given that, $w_{11} = 60$, $w_{12} = 80$, $w_{21} = 60$, $w_{22} = 80$, $w_{13} = -60$, $w_{23} = 60$, $b_1 = -90$, $b_2 = -40$ and $b_3 = -30$. Use the sigmoid transfer function in your program.

17.2 Edit Program_17c.py to update the weights w_{11}, w_{12}, w_{21} and w_{22}, for Example 17.2.1.

17.3 Use Python to investigate an ANN for the Boston housing data which has one hidden layer. How many neurons are required in the hidden layer to minimize the errors?

17.4 Given that $b_2 = -1$, $w_{11} = 1.5$, $w_{21} = 5$, $\alpha = 1$ and $\beta = 0.1$ plot bifurcation diagrams, with feedback, for system (17.4) when:

(i) $w_{12} = 3$, and $-5 \leq b_1 \leq 5$;

(ii) $w_{12} = -5$, and $-5 \leq b_1 \leq 5$;

(iii) $w_{12} = 8$, and $-5 \leq b_1 \leq 5$.

(iv) $b_1 = 1$, and $-10 \leq w_{12} \leq 10$.

How do the results compare with the stability diagram shown in Figure 17.7(b)?

Solutions to the Exercises may be found here:

`https://drstephenlynch.github.io/webpages/Solutions_Section_3.html`.

FURTHER READING

[1] Aggarwal, C.C. (2018). *Neural Networks and Deep Learning: A Textbook.* Springer, New York.

[2] Gad, A.F. and Jarmouni F.E. (2020). *Introduction to Deep Learning and Neural Networks with Python: A Practical Guide.* Academic Press, Cambridge, MA.

[3] Hebb, D.O. (1949). *The Organization of Behavior,* John Wiley, New York.

[4] Loy, J. (2020). *Neural Network Projects with Python: The ultimate guide to using Python to explore the true power of neural networks through six projects.* Packt Publishing, Birmingham, UK.

[5] McCulloch, W.S. and Pitts, W. (1943). A logical calculus of the ideas immanent in nervous activity. *Bull. of Math. Biophysics.* **5** 115–133.

[6] Minsky, M. and Papert, S. (1969). *Perceptrons.* MIT Press, Cambridge, MA.

[7] Rashid, T. (2016). *Make Your Own Neural Network.* Create Space Independent Publishing Platform, Scotts Valley, CA.

[8] Rosenblatt, F. (1958). The perceptron: A probabilistic model for information storage and organization in the brain. *Psychological Review.* **65** 386–408.

[9] Rumelhart, D.E. and McClelland, J.L. Eds. (1986). *Parallel Distributed Processing: Explorations in the Microstructure of Cognition.* MIT Press, Cambridge, MA.

[10] Taylory, M. (2017). *Make Your Own Neural Network: An In-depth Visual Introduction For Beginners.* Independently Published.

[11] Turing, A.M. (1937). On computable numbers, with an application to the entscheidungsproblem. *Proceedings of the London Mathematical Society.* **2**(42) 230–265.

[12] Turing, A.M. (1950). Computing machinery and intelligence. *Mind* **49** 433-460.

## CHAPTER 18

TensorFlow and Keras

TensorFlow is a Python library for heavy numerical computation in artificial intelligence (AI) written in Python, C++ and CUDA (C++ is a faster programming language than Python). The numpy library preprocesses data before it is fed in to TensorFlow, making it extremely efficient. TensorFlow uses tensors in the computations, you can think of these as arrays of arrays, as described in Section 2.1 of this book. TensorFlow was first released in 2015 by Google and the latest version 2, was released in 2019. It is predominantly used in machine and deep learning within AI and is completely free to the user. The easiest way to use TensorFlow is through Google Colab, and users have access to Central Processing Units (CPUs), Graphical Processing Units (GPUs), and even Tensor Processing Units (TPUs), via Google Colab and cloud computing. It is also possible to install TensorFlow to use in Spyder, readers will be able to find the latest instructions on the web. The URL for TensorFlow is

https://www.tensorflow.org,

where readers can find more information, examples at the beginner, intermediate and expert levels, and videos made by a diverse community of developers, on how it is being applied in the real world. Currently, the bestselling book on TensorFlow and Keras is [3], and other books include [2] and [5], for example.

Keras is an Application Programming Interface (API) adopted by TensorFlow. It provides a user-friendly interface where activation functions, cost functions, metrics, neural network layers, normalization schemes, and optimizers, for example, can all be incorporated in just a few lines of code. The URL for Keras is

https://keras.io.

Aims and Objectives:

- To introduce artificial intelligence.

- To introduce TensorFlow for machine and deep learning.

- To list some useful Keras commands.

DOI: 10.1201/9781003285816-18

At the end of the chapter, the reader should be able to

- use TensorFlow as a tool in AI;

- set up deep neural networks using Keras;

- interpret the results in real-world terms.

18.1   ARTIFICIAL INTELLIGENCE

Artificial intelligence (AI) refers to computational systems that behave in a similar manner to the human brain. AI systems date back to ancient times, in Greek mythology, Talos was a giant AI bronze statue built by the Gods to protect the island of Crete (it appears in one of my favorite movies—Jason and the Argonauts). More recently, Mary Shelley's Frankenstein, published in 1818, and Rossum's Universal Robots, a science fiction play, from 1921, both play on the fears that AI could lead to the destruction of the human race. To add to the negative historical portrayal of AI, in 1968, Arthur C. Clarke's book and the accompanying Stanley Kubrick movie, 2001: A Space Odyssey, depict AI and the computer HAL-9000 supercomputer in a negative way. Modern science is attempting to use AI for the benefit of the human race and it will permeate all of our lives for many years to come.

After the development of neural networks, covered in Chapter 17, the first real breakthrough for AI came in 1986, with the development of the backpropagation algorithm [8]. The next major breakthrough came in 1997, when IBM's Deep Blue computer beat the world chess champion, Gary Kasparov, under tournament conditions. Another major milestone followed in 2012, when Hinton and his group developed AlexNet for the ImageNet challenge in 2012, see [4], which was a major result in the field of computer vision. In 2014, Google-backed DeepMind Technologies learnt and successfully played forty nine classic Atari games by itself using deep reinforcement learning. In 2016, AlphaGo used a Monte-Carlo tree search algorithm to beat a Chinese champion. In natural language processing (NLP), there was a major breakthrough in 2018, with Google's BERT setting new standards, and in 2019, TensorFlow 2 was released. For further information on AI, readers are directed to [1], [6] and [7], for example.

AI can be broken down into a number of subcategories, including:

Computer Vision: Uses digital images from cameras and video cameras to identify and classify objects, and react accordingly to them. In this book, convolutional neural networks are used to identify objects in images.

Deep Learning: Is a subset of machine learning, where the networks are deeper and are modeled on structures of the human brain. Deep learning is used in Chapters 18, 19 and 20 of this book.

Machine Learning: Focuses on the use of algorithms and data to imitate the way the human brain learns. An example is presented in Section 16.3, where the values of houses in Boston districts can be predicted.

Natural Language Processing: Is the analysis and syntheses of natural language, speech and text, using computational techniques. The final chapter includes an example from Google Colab, where new text in the style of William Shakespeare is created.

Artificial Neural Networks (ANNs): Computing systems inspired by human brain dynamics. Chapter 17 provides an introduction to the subject. There are basically two subcategories:

Recurrent Neural Networks (RNNs): Are also covered in Chapter 19, and are network architectures for deep learning and use sequential or time series data.

Convolutional Neural Networks (CNNs): Are covered in Chapter 20, and are network architectures for deep learning with applications in image data and signal data.

18.2   LINEAR REGRESSION IN TENSORFLOW

As a simple introduction to TensorFlow consider the simplest neural network possible, consisting of one input, one bias, one neuron, and one output, see Figure 18.1(a).

Example 18.2.1. Use numpy to generate fifty random data points from linear data and use TensorFlow and Keras for linear regression and explain the results.

Solution. Program_18a.ipynb is split into three parts. The first part creates fifty random data points, labelled x_train and y_train, distributed around the line $y = x$, as shown in Figure 18.1(b). Next, the program uses Keras to create the ANN consisting of one layer with one neuron. The model is then compiled using a mean squared error and an Adam optimizer, which is a replacement algorithm for stochastic gradient descent that can handle sparse gradients on noisy problems. The complete data set is fed through 100 times, and the loss is used from the history metrics to plot a Cost versus Epochs plot, as shown in Figure 18.1(c). Finally, the learned data points are used to plot the line of best fit, as displayed in Figure 18.1(d). The weight w_1, gives the gradient of the line of best fit, and the bias is the y-intercept.

```
# Program_18a.ipynb: Simple Linear Regression with TensorFlow.
# The simplest way to use TensorFlow is through Google Colab.
import numpy as np
import tensorflow as tf
import matplotlib.pyplot as plt
np.random.seed(101)
```

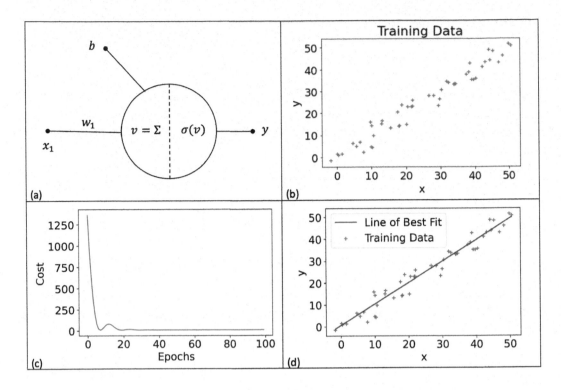

Figure 18.1 (a) A one-neuron perceptron ANN for linear regression. (b) The data points. (c) Cost against the number of epochs. The cost reduces to almost zero after just 30 epochs. (d) The line of best fit for the data.

```
# Generate the data.
x_train = np.linspace(0, 50, 50)
y_train = np.linspace(0, 50, 50)  # There will be 50 data points.
x_train += np.random.uniform(-4, 4, 50)    # Add noise to the data.
y_train += np.random.uniform(-4, 4, 50)
plt.rcParams["font.size"] = "18"
n = len(x_train)                           # Number of data points.
plt.scatter(x_train , y_train , marker = "+")
plt.xlabel("x")
plt.ylabel("y")
plt.title("Training Data")
plt.show()
# Set up the neural network.
layer0 = tf.keras.layers.Dense(units=1, input_shape=[1])
model = tf.keras.Sequential([layer0])
model.compile(loss='mean_squared_error',
              optimizer=tf.keras.optimizers.Adam(0.1))
history = model.fit(x_train, y_train, epochs=100, verbose=False)
plt.rcParams["font.size"] = "18"
```

```
plt.xlabel('Epochs')
plt.ylabel("Cost")
plt.plot(history.history["loss"])
plt.show()
# Display the results.
weights = layer0.get_weights()
weight = weights[0][0]
bias = weights[1]
print('weight: {} bias: {}'.format(weight, bias))
y_learned = x_train * weight + bias
plt.rcParams["font.size"] = "18"
plt.scatter(x_train, y_train, marker = "+" , label = "Training Data")
plt.plot(x_train,y_learned,color="red",label="Line of Best Fit")
plt.legend()
plt.xlabel("x")
plt.ylabel("y")
plt.show()
```

18.3   XOR LOGIC GATE IN TENSORFLOW

An ANN for the XOR logic gate is shown in Figure 16.4(a) and the corresponding truth table is given in Figure 16.4(b). The second simple example of the chapter shows how to use TensorFlow to train an ANN for an XOR logic gate.

Example 18.3.1. Use TensorFlow and Keras to train an ANN for the XOR logic gate. Recall that this problem is not linearly separable, so you need more than one layer in the ANN.

Solution. Program_18b.ipynb lists the Python code for the ANN for the XOR logic gate. The necessary libraries are loaded and the training data and target data are set. Next, Keras is used to set up the ANN, we have a sequential model which is a linear stack of layers. There are eight neurons in layer one, eight in layer two, and one in the output layer. This configuration gives good results for relatively low numbers of epochs. A new activation function is introduced in this program, namely the rectified linear activation function (ReLU) and the sigmoid activation function is used for the output neuron. The ReLU activation function is a piecewise linear function and its graph is displayed in Figure 18.2(a). For completeness, the graphs of three other activation functions defined in Keras are included in Figure 18.2. The reader will be asked to plot these graphs and their derivatives in the exercises at the end of the chapter.

```
# Program_18b.ipynb: TensorFlow for the XOR Logic Gate.
import numpy as np
import matplotlib.pyplot as plt
import tensorflow as tf
from tensorflow import keras
```

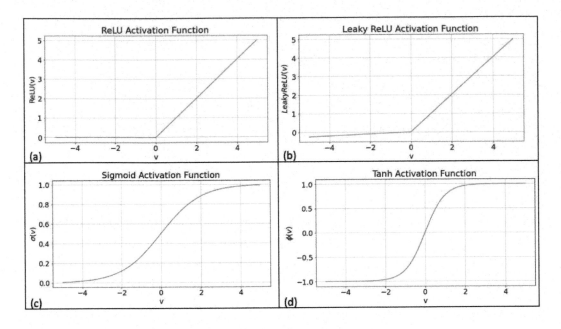

Figure 18.2 Some activation functions used by Keras: (a) ReLU. (b) Leaky ReLU. (c) Sigmoid, see equation (17.2). (d) Tanh, see equation (17.3).

```
training_data = np.array([[0,0] , [0,1] , [1,0] , [1, 1]] , "float32")
target_data = np.array([[0] , [1] , [1] , [0]] , "float32")
model = tf.keras.models.Sequential()
model.add(tf.keras.layers.Dense(8 , activation = "relu"))
model.add(tf.keras.layers.Dense(8 , activation = "relu"))
model.add(tf.keras.layers.Dense(1 , activation = "sigmoid"))
model.compile(loss="mean_squared_error", \
              optimizer=tf.keras.optimizers.Adam(0.1), \
              metrics=["accuracy"])
hist = model.fit(training_data,target_data,epochs=500,verbose=1)
print(model.predict(training_data).round())
val_loss, val_acc = model.evaluate(training_data, target_data)
print(val_loss, val_acc)
loss_curve = hist.history["loss"]
acc_curve = hist.history["accuracy"]
plt.plot(loss_curve , label="Loss")
plt.plot(acc_curve , label="Accuracy")
plt.xlabel("Epochs")
plt.legend()
plt.show()

for layer in model.layers: \
print(layer.get_config(), layer.get_weights())
```

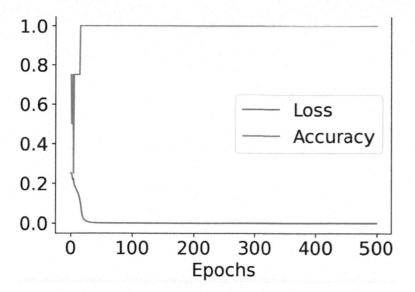

Figure 18.3 Output from Program_18b.py, the loss and accuracy curves.

The program is run over 500 epochs and the output is a plot of the accuracy and loss curves, see Figure 18.3. Setting **verbose=1**, means that the output is printed on every epoch. To switch this off simply set **verbose=0**. The final command line prints out the values of the biases and weights for all of the layers. The reader will be asked to investigate the configuration of the ANN in the exercises at the end of the chapter. Figure 16.4(a) shows that one hidden layer with two neurons, and one output neuron is sufficient for the XOR logic gate problem, however, it may not be the most efficient ANN.

18.4   BOSTON HOUSING DATA IN TENSORFLOW AND KERAS

The Boston housing data was introduced in Chapter 17 when discussing the perceptron and machine learning. In this section, TensorFlow is used to build more complex ANNs, and it is shown that this generally leads to overfitting for this particular problem.

Example 18.4.1. Use TensorFlow and Keras on the Boston housing data to see how the structure of an ANN affects the results.

Solution. Program_18c.ipynb lists four separate models for the Boston housing data. in the first case, one neuron is used, as in Example 16.3.1, there are 1000 epochs, and the validation split is 0.2, meaning that 80% of the data is used to train the network, and 20% is used to test the network. Figure 18.4(a) shows that there is good agreement between the training data and the test data. For the second model, ten neurons are introduced in a hidden layer and the validation split remains at 20%. The results are displayed in Figure 18.4(b), and there is clearly a split between the loss curves, indicating some overfitting. For the third model, 100 neurons are introduced

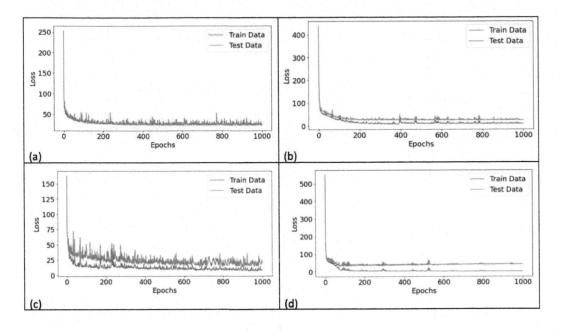

Figure 18.4 Using TensorFlow and Keras on ANNs for the full Boston housing data, loss curves for training and test data: (a) Model 1: One neuron and 20% of the data is used to test. There is no overfitting. (b) Model 2: Ten neurons in a hidden layer and 20% of the data is used to test. There is some overfitting as the loss curves are separated. (c) Model 3: One hundred neurons in each of two hidden layers and 20% of the data is used to test. There is some overfitting. (d) Model 4: One hundred neurons in each of two hidden layers and 90% of the data is used to test. This is the worst set of results with the test data curve moving away from the training data curve.

in hidden layer one and 100 neurons are inserted in hidden layer 2. The validation split remains at 20%. The results are displayed in Figure 18.4(c), and there is clearly a split between the loss curves, indicating some overfitting. Finally, in model 4, the same ANN is used as in model 3, however, the validation split is set at 90%. Figure 18.4(d), shows the worst set of results, where the curve for the validation data is actually moving away from the curve for the training data.

```
# Program_18c.ipynb: TensorFlow programs for Boston housing data.
import tensorflow as tf
from tensorflow import keras
import numpy as np
import matplotlib.pyplot as plt
# Import the data through Keras.
from keras.datasets import boston_housing
(x_train, y_train), (x_test, y_test)=boston_housing.load_data(path= \
"boston_housing.npz",test_split=0,seed=113)
# Model 1: One neuron, no overfitting.
model = keras.Sequential([keras.layers.Dense(1, input_dim=13, \
```

```
kernel_initializer="normal"),])
# Model 2: Overfitting due to too many neurons.
# model = keras.Sequential([
# keras.layers.Dense(10, input_dim=13, kernel_initializer="normal", \
# activation="relu"),
# keras.layers.Dense(1, input_dim=13, kernel_initializer="normal"),])
# Model 3: Overfitting due to too many layers and neurons.
# model = keras.Sequential([
# keras.layers.Dense(100, input_dim=13, kernel_initializer="normal", \
# activation="relu"),
# keras.layers.Dense(100, kernel_initializer="normal", \
# activation="relu"),
# keras.layers.Dense(1, input_dim=13, kernel_initializer="normal"),])
# Model 4: Validation split: 90% of data used to test.
# Change validation_split to 0.9 in hist.
# model = keras.Sequential([
# keras.layers.Dense(100, input_dim=13, kernel_initializer="normal", \
# activation="relu"),
# keras.layers.Dense(100, kernel_initializer="normal", \
# activation="relu"),
# keras.layers.Dense(1, input_dim=13, kernel_initializer="normal"),])
model.compile(loss="mean_squared_error", optimizer= \
tf.keras.optimizers.Adam(0.01))
hist=model.fit(x_train, y_train, epochs=1000, validation_split=0.2, \
verbose=0)
epochs = 1000
plt.rcParams["font.size"] = "18"
plt.plot(range(epochs), hist.history["loss"], range(epochs), \
hist.history["val_loss"])
plt.xlabel("Number of Epochs")
plt.ylabel("Loss")
plt.show()
```

Keras is an API designed for human beings, it is relatively simple to use, flexible and incredibly powerful. Table 18.1 lists some Keras API features. Readers are asked to go through each item in the exercises at the end of the chapter.

Within the Keras web pages readers will find plenty of examples in computer vision, NLP, structured data, time series, audio data, generative deep learning, reinforcement learning, graph data and quick Keras recipes. Users of Keras are also encouraged to add their own examples.

EXERCISES

18.1 Create simple TensorFlow programs for the logical AND and OR gate ANNs introduced in Chapter 16.

18.2 Look on the internet for definitions of the ReLU and leaky ReLU activation functions. Use Python to plot the activation functions shown in Figure 18.2. In each case, plot the corresponding derivative curve on the same graph.

18.3 Create a TensorFlow program for the ANN for the XOR gate shown in Figure 17.4(a). There are two inputs, two neurons in the hidden layer, and one output neuron. Print out the weights of the trained network and investigate how the Adam learning rate and numbers of epochs affects the accuracy.

18.4 Investigate each item from the Keras API reference list shown in Table 18.1. The reader is also encouraged to run through the examples provided within the Keras web pages.

Table 18.1   Keras API Reference, see **https://keras.io**

Models API	Losses	Data Loading
Model Class Sequential class Model training APIs Model saving, serialization	Probabilistic Regression Hinge	Image data Timeseries data Textdata
Layers API	**Callbacks API**	**Smart Datasets**
Base layer class Layer activations Layer weight initializers Layer weight Regularizers Core layers Convolution layers Pooling layers Recurrent layers Preprocessing layers Normalization layers Regularization layers Attention layers Reshaping layers Merging layers Locally-connected layers Activation layers	Base callback class Model checkpoint TensorBoard Early stopping Learning rate scheduler Reduce LR on plateau Remote monitor Lambda callback Terminate on NaN CSV logger Progbar logger Back up and restore	MNIST digits CIFAR10 small images CIFAR100 small images IMDB movie review Reuters newswire Fashion MNIST Boston housing **Utilities** Model plotting Serialization Python and numpy Backend
Optimizers	**Metrics**	**Mixed Precision**
SGD RMSprop Adam Adadelta Adagrad Adamax Nadam Ftrl	Accuracy Probabilistic Regression Classification T/F Image segmentation Hinge (max margin)	Policy API Loss scale optimizer

Solutions to the Exercises may be found here:

`https://drstephenlynch.github.io/webpages/Solutions_Section_3.html.`

FURTHER READING

[1] Aggarwal, C.C. (2021). *Artificial Intelligence: A Textbook.* Springer, New York.

[2] Audevart, A., Banachewicz, K. and Massaron, L. (2021). *Machine Learning Using TensorFlow Cookbook: Create powerful machine learning algorithms with TensorFlow.* Packt Publishing, Birmingham.

[3] Geron, A. (2019). *Hands-On Machine Learning with Scikit-Learn, Keras, and Tensor-Flow: Concepts, Tools, and Techniques to Build Intelligent Systems, 2nd Ed.* O'Reilly Media, Sebastopol, CA.

[4] Krizhevsky, A., Sutskever, I. and Hinton, G.E. (2012). ImageNet classification with deep convolutional neural networks. *Nips'12. Curran Associates Inc.* 1097–1105.

[5] Moocarme, M. So, A. and Maddalone, A. (2021). *The TensorFlow Workshop: A hands-on guide to building deep learning models from scratch using real-world datasets.* Packt Publishing, Birmingham.

[6] Moroney, L. (2020). *AI and Machine Learning for Coders: A Programmer's Guide to Artificial Intelligence.* O'Reilly Media, Sebastopol, CA.

[7] Norvig, P. and Russell, S. (2021). *Artificial Intelligence: A Modern Approach, Global Edition, 4th Ed.* Pearson, London.

[8] Rumelhart, D.E., Hinton, G.E. and Williams, R.J. (1986). Learning representations by back-propagating errors. *Nature* **323**, 533–536.

## CHAPTER 19

Recurrent Neural Networks

Recurrent neural networks (RNNs) are a type of artificial neural network (ANN) which take information from previous inputs which in turn influences current input and output. In this way, RNNs feed back into themselves in a cyclic fashion and have an internal memory. They are used for predicting time series, music composition, machine translation, and natural language processing, including speech recognition and sentiment analysis, for example. RNNs are employed in Apple's Siri, Google's voice search, and Google translate. For more information, the reader is guided to [2], [10] and [15], for example.

The chapter starts with an introduction to the discrete Hopfield RNN, first used as an addressable memory in the 1980s. The second section covers continuous Hopfield models and a theorem is employed to compute a Lyapunov function to determine stability conditions. The now famous long short-term memory (LSTM) architecture was invented in 1997, and is used in this book to predict chaotic and financial time series in the third and fourth sections, respectively.

Aims and Objectives:

- To introduce discrete and continuous Hopfield RNNs.

- To describe LSTM RNNs.

At the end of the chapter, the reader should be able to

- use discrete Hopfield networks as addressable memories;

- determine Lyapunov functions for Hopfield networks and plot surface and contour plots;

- use LSTM neural networks to predict time series data of chaotic systems;

- use LSTM RNNs to predict financial time series.

19.1 THE DISCRETE HOPFIELD RNN

In 1982, Hopfield [6] used his network as a content-addressable memory RNN using fixed points as attractors for certain fundamental memories. A five-neuron Hopfield

DOI: 10.1201/9781003285816-19

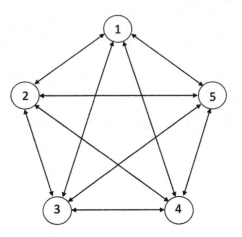

Figure 19.1 A five-neuron Hopfield RNN used as an addressable memory. The synaptic weights between neurons i and j are $w_{ij} = w_{ji}$, respectively.

RNN is displayed in Figure 19.1. Consider an n-neuron Hopfield model which can be implemented using the following four-step algorithm.

1. **Hebb's Postulate of Learning.** Let $\mathbf{x}_1, \mathbf{x}_2, \ldots, \mathbf{x}_M$ denote a set of N-dimensional fundamental memories. The synaptic weights of the network are determined using the equation

$$W = \frac{1}{N} \sum_{r=1}^{M} \mathbf{x}_r \mathbf{x}_r^T - \frac{M}{N} \mathbf{I}_n, \tag{19.1}$$

   where $\mathbf{I}_n$ is an identity matrix. Once computed, the synaptic weights w_{ij}, say, remain fixed.

2. **Initialization.** Let $\mathbf{x}_p$ denote the unknown probe vector to be tested. The algorithm is initialized by setting

$$x_i(0) = x_{ip}, \quad i = 1, 2, \ldots, N,$$

   where $x_i(0)$ is the state of neuron i at time $n = 0$, x_{ip} is the ith element of vector $\mathbf{x}_p$, and N is the number of neurons.

3. **Iteration.** The elements are updated asynchronously (i.e., one at a time in a random order) according to the rule

$$x_i(n+1) = \text{hsgn}\left(\sum_{j=1}^{N} w_{ij} x_j(n) \right), i = 1, 2, \ldots, N,$$

   where

$$\text{hsgn}(v_i(n+1)) = \begin{cases} 1, & v_i(n+1) > 0 \\ x_i(n), & v_i(n+1) = 0 \\ -1, & v_i(n+1) < 0 \end{cases}$$

and $v_i(n + 1) = \sum_{j=1}^{N} w_{ij}x_j(n)$. The iterations are repeated until the vector converges to a stable value. Note that at least N iterations are carried out to guarantee convergence.

4. **Result.** The stable vector, say, $\mathbf{x}_{\text{fixed}}$, is the result.

The weights are updated asynchronously one at a time. Synchronous updating is the procedure by which weights are updated simultaneously. The fundamental memories should first be presented to the Hopfield network. This tests the network's ability to recover the stored vectors using the computed synaptic weight matrix. The desired patterns should be recovered after one iteration; if not, then an error has been made. Distorted patterns or patterns missing information can then be tested using the above algorithm. There are two possible outcomes.

1. The network converges to one of the fundamental memories.

2. The network converges to a *spurious steady-state*. Spurious steady-states include the following:

   (a) *Reversed fundamental memories*—e.g., if $\mathbf{x}_f$ is a fundamental memory then so is $-\mathbf{x}_f$.

   (b) *Mixed fundamental memories*—a linear combination of fundamental memories.

   (c) *Spin-glass states*—local minima not correlated with any fundamental memories.

Example 19.1.1. A five-neuron discrete Hopfield network is required to store the following three fundamental memories:

$$\mathbf{x}_1 = (1, 1, 1, 1, 1)^T, \quad \mathbf{x}_2 = (1, -1, -1, 1, -1)^T, \quad \mathbf{x}_3 = (-1, 1, -1, 1, 1)^T.$$

Write a Python program to:

(a) Compute the synaptic weight matrix W.

(b) Use asynchronous updating to show that the three fundamental memories are stable.

(c) Test the following vectors on the Hopfield network (the random orders affect the outcome):

$$\mathbf{x}_4 = (1, -1, 1, 1, 1)^T, \quad \mathbf{x}_5 = (0, 1, -1, 1, 1)^T, \quad \mathbf{x}_6 = (-1, 1, 1, 1, -1)^T.$$

Solution. The program for computing the weight matrix W, see equation (19.1), and testing vectors is listed below. The three fundamental memories are stable, for example, inputting vector $\mathbf{x}_1$ will return the vector $\mathbf{x}_1$. Run the program a few times for vectors $\mathbf{x}_4$ to $\mathbf{x}_6$. Readers are also encouraged to compute the weight matrix, W, and output vectors by hand, choosing random ordering from your head.

```
# Program_19a.py: The Hopfield network used as a memory.
from sympy import Matrix, eye
import random
# The fundamental memories:
x1,x2,x3 = [1, 1, 1, 1, 1] , [1, -1, -1, 1, -1] , [-1, 1, -1, 1, 1]
X = Matrix([x1, x2, x3])
W = X.T * X / 5 - 3*eye(5) / 5
print("W = " , W)
# The hsgn function.
def hsgn(v, x):
    if v > 0:
        return 1
    elif v == 0:
        return x
    else:
        return -1
# Choose the elements at random.
L = [0, 1, 2, 3, 4]
n = random.sample(L, len(L))
# Input vector.
xinput = [1, -1, -1, 1, 1]
xtest = xinput
for j in range(4):
    M = W.row(n[j]) * Matrix(xtest)
    xtest[n[j]] = hsgn(M[0], xtest[n[j]])
if xtest == x1:
    print('Net has converged to X1')
elif xtest == x2:
    print('Net has converged to X2')
elif xtest == x3:
    print('Net has converged to X3')
else:
    print('Iterate again: May have converged to spurious state')
```

19.2   THE CONTINUOUS HOPFIELD RNN

In 1984, Hopfield introduced the continuous Hopfield RNN [7] and it was demonstrated how an analog electrical circuit could behave as a small network of neurons with graded response. Hopfield derived a so-called Lyapunov function, used to check for stability, and employed the network as a content addressable memory. For more information on Lyapunov functions and stability, see [13], for example. Nowadays, these RNNs can be used to solve some classification and optimization problems, where the critical points of the dynamical system are the solutions to the problem. In [8], for example, Hopfield used his networks to solve the travelling salesman problem, and in 2017, a divide-and-conquer strategy was adopted to improve the model [4]. Using Kirchoff's circuit laws, Hopfield derived the following differential equations:

$$\frac{d}{dt}\mathbf{x}(t) = -\mathbf{x}(t) + W\mathbf{a}(t) + \mathbf{b}, \tag{19.2}$$

where $\mathbf{x}(t)$ is a vector of neuron activation levels, W is the weight matrix, representing synaptic connections, $\mathbf{b}$ are the biases, $\mathbf{a}(t) = \phi(\mathbf{x}(t))$ are the nonlinear input/output activation levels, and ϕ is an activation function. Hopfield derived the following theorem for stability properties.

Theorem 19.1. *A Lyapunov function for the n-neuron Hopfield network defined by equation (19.2) is given by*

$$V(\mathbf{a}) = -\frac{1}{2}\mathbf{a}^T W\mathbf{a} + \sum_{i=1}^{n}\left(\int_0^{a_i}\phi^{-1}(u)du\right) - \mathbf{b}^T\mathbf{a} \tag{19.3}$$

as long as

1. *$\phi^{-1}(a_i)$ is an increasing function, that is,*

$$\frac{d}{da_i}\phi^{-1}(a_i) > 0, \quad and$$

2. *the weight matrix W is symmetric.*

Consider the following two-neuron example.

Example 19.2.1. A schematic of a two-neuron Hopfield RNN is shown in Figure 19.2. Suppose that

$$\frac{dx}{dt} = -x + 2\left(\frac{2}{\pi}\tan^{-1}\left(\frac{\gamma\pi x}{2}\right)\right), \quad \frac{dy}{dt} = -y + 2\left(\frac{2}{\pi}\tan^{-1}\left(\frac{\gamma\pi y}{2}\right)\right). \tag{19.4}$$

Use Hopfield's theorem to determine a suitable Lyapunov function for system (19.4) and plot a surface and contour plot of this function when $\gamma = 0.7$. How many stable states are there?

Solution. From equation (19.4),

$$W = \begin{pmatrix} 2 & 0 \\ 0 & 2 \end{pmatrix}, \quad \mathbf{b} = \begin{pmatrix} 0 \\ 0 \end{pmatrix}, \quad a_1(t) = \frac{2}{\pi}\tan^{-1}\left(\frac{\gamma\pi x}{2}\right), \quad a_2(t) = \frac{2}{\pi}\tan^{-1}\left(\frac{\gamma\pi y}{2}\right).$$

A Lyapunov function, derived using equation (19.3), is given by

$$V(\mathbf{a}) = -\frac{1}{2}(a_1 \ a_2)\begin{pmatrix} 2 & 0 \\ 0 & 2 \end{pmatrix}\begin{pmatrix} a_1 \\ a_2 \end{pmatrix} + \int_0^{a_1}\phi^{-1}(u)du +$$

$$\int_0^{a_2}\phi^{-1}(u)du - (0 \ 0)\begin{pmatrix} a_1 \\ a_2 \end{pmatrix}.$$

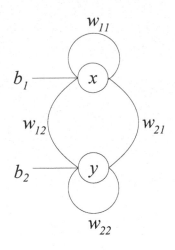

Figure 19.2 A two-neuron Hopfield network, where x and y denote the activation levels of the neurons.

Therefore,

$$V(a_1, a_2) = -\left(a_1^2 + a_2^2\right) - \frac{4}{\gamma\pi^2}\left(\log\left(\cos(\pi a_1/2)\right) + \log\left(\cos(\pi a_2/2)\right)\right). \qquad (19.5)$$

Program_19b.py produces the contour plot in Figure 19.3(a) and the surface plot in Figure 19.3(b). Note that the V-range is, $-0.2 \leq V \leq 0$, there is one local maximum and there are four local minima, indicating four stable points for the Hopfield RNN.

```
# Program_19b.py: Contour and surface plots of the Hopfield Lyapunov
# function. Run in Spyder, in the Console window, type >>> matplotlib
# qt5. You can then rotate the 3D figures.
import numpy as np
import matplotlib.pyplot as plt
from mpl_toolkits import mplot3d
gamma = 0.7
def flux_qubit_potential(x,y):
    return -(x**2+y**2)- (4/(gamma*np.pi**2)) * \
    (np.log(np.cos(np.pi*x/2)) + np.log(np.cos(np.pi*y/2)))
y=np.linspace(-1,1,60)
x=np.linspace(-1,1,60)
X,Y=np.meshgrid(x,y)
Z=flux_qubit_potential(X,Y).T

fig=plt.figure(figsize = (10,10))
plt.subplot(1,2,1)
ax=fig.add_subplot(1,2,1,projection='3d')
p=ax.plot_wireframe(X, Y, Z, rstride=3, cstride=3)
ax.plot_surface(X,Y,Z,rstride=3,cstride=3,alpha=0.1)
```

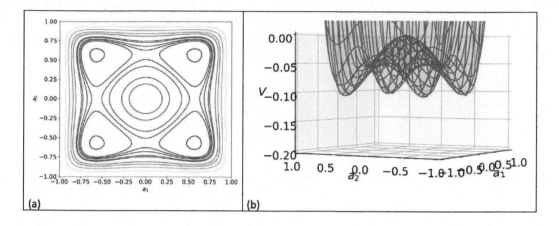

Figure 19.3 (a) A contour plot of the Lyapunov function (19.5). (b) The corresponding surface plot wireframe, which has one local maximum and four local minima. A wireframe is used in order to see through the surface. Readers can rotate the three-dimensional image in Spyder.

```
ax.set_xlim3d(-1, 1);
ax.set_ylim3d(-1, 1);
ax.set_zlim3d(-0.2, 0);
ax.set_xlabel("$a_1$", fontsize = 15)
ax.set_ylabel("$a_2$", fontsize = 15)
ax.set_zlabel("$V$", fontsize = 15)
plt.tick_params(labelsize = 15)
plt.show()
plt.subplot(1,2,2)
plt.contour(X,Y,Z,[-0.1,-0.06,-0.05,-0.04,-0.02,-0.01,0.05])
plt.xlabel(r'$a_1$',fontsize=15)
plt.ylabel(r'$a_2$',fontsize=15)
plt.tick_params(labelsize=15)
plt.show()
```

19.3 LSTM RNN TO PREDICT CHAOTIC TIME SERIES

In 1997, Hochreiter and Schmidhuber invented the LSTM architecture [5] which partially solved the vanishing gradient problem when using backpropagation in RNNs. Figure 19.4(a) shows a single-unit RNN, where the output of a particular layer is fed back to the input to predict the output layer. The middle layer can consist of multiple hidden layers as shown in Figure 19.4(b), which shows the configuration of a LSTM unit. The data in a LSTM unit can be stored, written, or read, by means of gates that open and close. The sigmoid functions maintain values between 0 and 1, and help the network update or forget the data. The tanh functions regulate data flowing through the network and are used to avoid information fading. For more details, see [5].

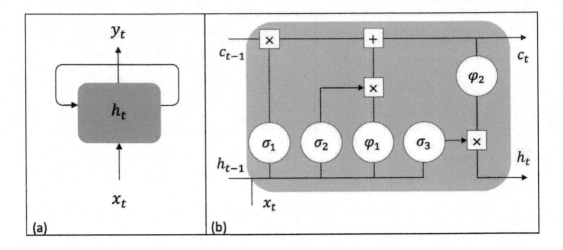

Figure 19.4 (a) A single-unit RNN, where x_t, h_t, y_t are the input, hidden, and output states, respectively. (b) A LSTM recurrent unit, where c_t, h_t are the cell and hidden states, respectively. The $\sigma_1, \sigma_2, \sigma_3$ neurons act as forget, input, and output gates, respectively. The ϕ_1, ϕ_2 neurons act as a candidate for cell state update, and updated cell state, to help determine the new hidden state, respectively. Note that $\sigma_{1,2,3}$ are sigmoid functions and $\phi_{1,2}$ are tanh functions.

Example 19.3.1. Use Keras, TensorFlow, and LSTM units to predict chaotic time series generated by the logistic map function, $f_\mu(x) = \mu x(1-x)$, when $\mu = 4$.

Solution. Program_19c.ipynb lists the code for producing the plots in Figures 19.5 to 19.8. Figure 19.5 shows the chaotic time series, recall that chaos is defined as random, unpredictable behavior. Figure 19.6 shows the loss curve for the 80% training data (blue curve), and the validation loss curve for the 20% testing data (orange curve). The RNN performs very well after about 28 epochs, and there is no evidence of overfitting. Figure 19.7 shows the historical, true and predicted values of the chaotic time series. Finally, Figure 19.8 is a zoom-in of Figure 19.7. The results are remarkable. The obvious question then is: Can this be applied to real-world data? The next section shows that indeed it can!

```
# Program_19c.ipynb: Predicting chaotic time series using a LSTM
# RNN, Keras, and TensorFlow.
import tensorflow as tf
from tensorflow import keras
import pandas as pd
import numpy as np
import seaborn as sns
from pylab import rcParams
import matplotlib.pyplot as plt
```

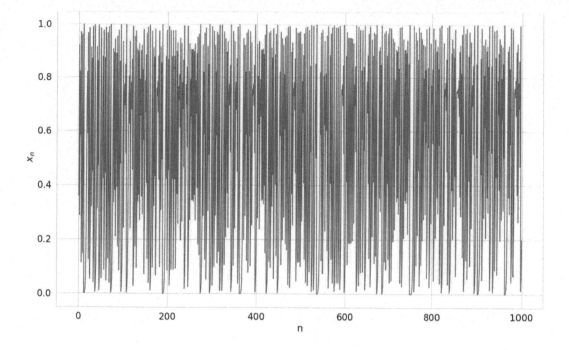

Figure 19.5 Time series of chaotic values generated by $x_{n+1} = 4x_n (1 - x_n)$, using an initial value $x_0 = 0.1$.

```
from matplotlib import rc
# Set up the figure.
%matplotlib inline
%config InlineBackend.figure_format='retina'
sns.set(style='whitegrid', palette='muted', font_scale=1.5)
rcParams['figure.figsize'] = 16, 10
# Create a chaotic time series using the logistic map function.
```

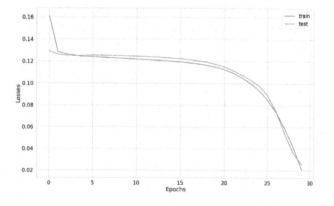

Figure 19.6 The loss curve for the training data (80% of the data), and the validation loss curve for the test data (20% of the data).

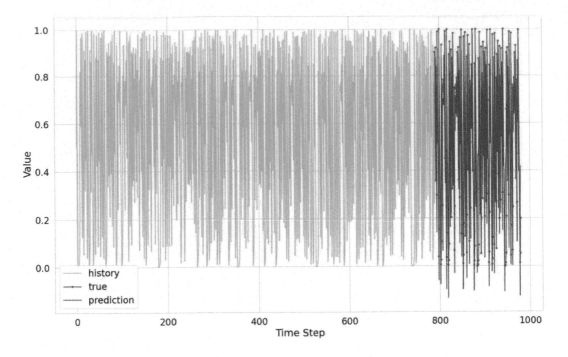

Figure 19.7 The green time series is the first 800 iterates, the blue times series, represents the true iterative values, and the red time series shows the predicted results.

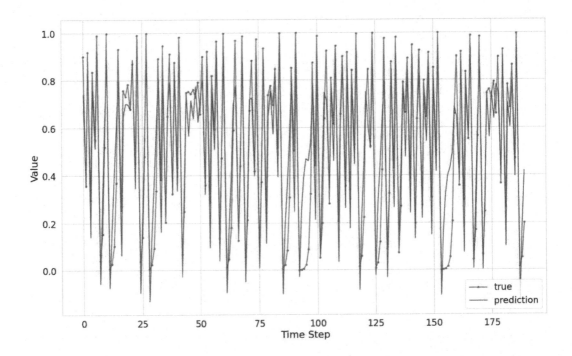

Figure 19.8 A zoom-in of Figure 19.7, showing the predicted and true time series.

```python
RANDOM_SEED = 42
np.random.seed(RANDOM_SEED)
x = 0.1
chaos = []
for t in range(1000):
  x = 4 * x * (1 - x)
  chaos = np.append(chaos,x)
time = np.arange(0, 100, 0.1)
plt.plot(chaos)
plt.xlabel("n")
plt.ylabel("$x_n$")
plt.show()
# Set up a data frame of the time series data.
df = pd.DataFrame(dict(chaos=chaos), index=time, columns=['chaos'])
df.head()
train_size = int(len(df) * 0.8)
test_size = len(df) - train_size
train, test = df.iloc[0:train_size], df.iloc[train_size:len(df)]
print(len(train), len(test))
def create_dataset(X, y, time_steps=1):
    Xs, ys = [], []
    for i in range(len(X) - time_steps):
        v = X.iloc[i:(i + time_steps)].values
        Xs.append(v)
        ys.append(y.iloc[i + time_steps])
    return np.array(Xs), np.array(ys)
time_steps = 10
# Reshape to [samples, time_steps, n_features]
X_train, y_train = create_dataset(train, train.chaos, time_steps)
X_test, y_test = create_dataset(test, test.chaos, time_steps)
print(X_train.shape, y_train.shape)
# Set up the RNN using Keras and TensorFlow.
model = keras.Sequential()
model.add(keras.layers.LSTM(128, input_shape=(X_train.shape[1], \
                X_train.shape[2])))
model.add(keras.layers.Dense(1))
model.compile(loss='mean_squared_error', \
                        optimizer=keras.optimizers.Adam(0.001))
history = model.fit(X_train, y_train, epochs=30, batch_size=16,
                    validation_split=0.1, verbose=0, shuffle=False)
plt.plot(history.history['loss'], label='train')
plt.plot(history.history['val_loss'], label='test')
plt.xlabel("Epochs")
plt.ylabel("Losses")
plt.legend()
```

```
plt.show()
# Plot the predicted and actual values.
y_pred = model.predict(X_test)
plt.plot(np.arange(0, len(y_train)), y_train, 'g', label="history")
plt.plot(np.arange(len(y_train), len(y_train) + len(y_test)), \
                        y_test, marker='.', label="true")
plt.plot(np.arange(len(y_train), len(y_train) + len(y_test)), \
                y_pred, 'r', label="prediction")
plt.ylabel('Value')
plt.xlabel('Time Step')
plt.legend()
plt.show()
# Show the predicted and actual values for the last 200 iterates.
plt.plot(y_test, marker='.', label="true")
plt.plot(y_pred, 'r', label="prediction")
plt.ylabel('Value')
plt.xlabel('Time Step')
plt.legend()
plt.show()
```

19.4   LSTM RNN TO PREDICT FINANCIAL TIME SERIES

In 2021, Korstanje published a book on advanced forecasting with Python including LSTM RNNs [9]. A number of journal papers have also been published, see, for example [1], [3] and [14]. The example in this section uses Keras, TensorFlow and LSTM RNNs to predict the US/EUR exchange rate.

Example 19.4.1. Use a LSTM RNN to predict the US/EUR exchange rates for data between September 2017 and September 2020.

Solution. Program_19d.ipynb lists the code and Figures 19.9 to 19.11 display the results. Readers can draw their own conclusions on the algorithms efficacy. Readers are encouraged to run models on other financial time series data sets.

```
# Program_19d.ipynb: Using LSTM to predict financial time series.
# of the US Dollar and Euro exchange rates.
# Importing the relevant libraries and API.
import pandas as pd
pd.options.display.max_colwidth = 60
!pip install fredapi
import tensorflow as tf
from tensorflow import keras
import numpy as np
from numpy import mean
```

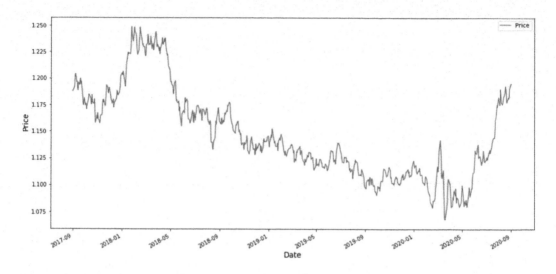

Figure 19.9 Time series data for the US/EUR exchange rate series between September 2017 and September 2020.

```
from fredapi import Fred          # Import the FRED API
import numpy as np
import math
import statistics
import pylab
%matplotlib inline
import matplotlib.pyplot as plt
```

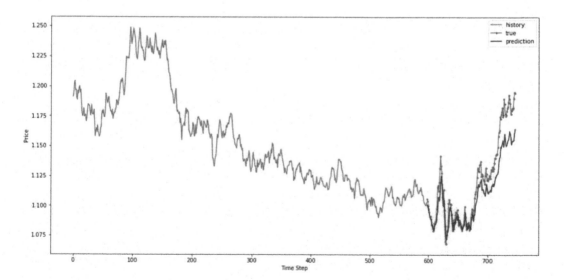

Figure 19.10 The blue time series is the actual data and the red time series is the predicted values for the last 20% of the data.

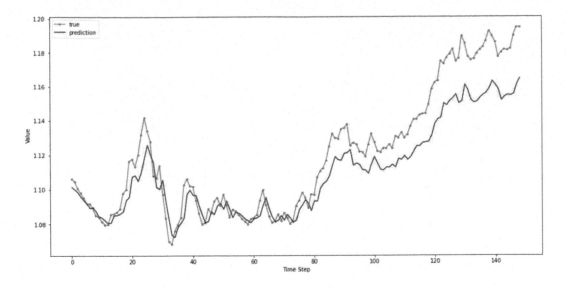

Figure 19.11 A zoom-in of the predicted data. The results appear to be diverging toward the end of the prediction.

```python
from IPython.core.pylabtools import figsize
figsize(16, 8)
fred = Fred(api_key='419401c4103800a51c50e6113b1f7500')
# Retrieve historical data for the US/EUR exchange rate series between
# Sept 2017 and Sept 2020.
data = fred.get_series('DEXUSEU',observation_start='2017-09-01', \
          observation_end='2020-09-01',name='Price')
data.head()
# Converting the data series into a dataframe.
Input = data.to_frame()
Input = Input.dropna()
plt.figure(1)
# Cleaning the input by replacing missing values.
clean_input= Input.fillna(method='ffill')
DATA = clean_input
# Renaming the price column from '0' to 'Price.
DATA.rename( columns={0 :'Price'}, inplace=True )
DATA.to_csv("EU_Dollar_Data.csv")
# Plotting the data.
DATA.plot()
plt.xlabel('Date',fontsize=14)
plt.ylabel('Price',fontsize=14)
plt.show()
# Split training and test data as 80-20 percent, respectively.
# Getting the number of rows to train the model on.
train_size= int(len(DATA)*0.8)
```

```
test_size= len(DATA) - train_size
# Splitting train and test data then printing the size (rows) of each.
train, test = DATA.iloc[0:train_size], DATA.iloc[train_size:len(DATA)]
print(len(train),len(test))
# Function:create_dataset.
# Converts data into numpy arrays.
def create_dataset(X,y,time_steps=1):
    Xs, ys= [],[]
    for i in range(len(X)-time_steps):
        v= X.iloc[i:(i+time_steps)].values
        Xs.append(v)
        ys.append(y.iloc[i+time_steps])
    return np.array(Xs), np.array(ys)
time_steps = 1
# Split data into X_train and y_train datasets.
X_train, y_train = create_dataset(train, train.Price, time_steps)
# Splitting test data into X_test and y_test datasets.
X_test, y_test = create_dataset(test, test.Price, time_steps)
print(X_train.shape, y_train.shape)
# Defining the LSTM network architecture.
model=keras.Sequential( )
model.add(keras.layers.LSTM(128,input_shape = \
                (X_train.shape[1],X_train.shape[2])))
model.add(keras.layers.Dense(1))
# Compile the model.
model.compile(loss='mean_squared_error',optimizer = \
                    keras.optimizers.Adam(0.0005))
history= model.fit(X_train, y_train, epochs=250, batch_size = 16,
                validation_split = 0.2 , verbose = 0, shuffle = False)
plt.figure(2)
plt.plot(history.history['loss'], label= 'train')
plt.plot(history.history['val_loss'], label= 'test')
plt.xlabel('Iterations')
plt.ylabel('Loss')
plt.legend()
plt.show()
plt.figure(3)
# Get the models predicted price values.
y_pred=model.predict(X_test)
# Plot the predictions along with the true outcomes
plt.plot(np.arange(0,len(y_train)), y_train, 'g',label="history")
plt.plot(np.arange(len(y_train), len(y_train)+ len(y_test)), y_test, \
            marker='.',label="true")
plt.plot(np.arange(len(y_train),len(y_train)+len(y_test)),y_pred,'r', \
            label="prediction")
```

Figure 19.12 The patterns to be used as fundamental memories for the discrete Hopfield model.

```
plt.ylabel('Price')
plt.xlabel('Time Step')
plt.legend()
plt.show()
# Zoom-in on predicted values.
plt.figure(4)
plt.plot(y_test, marker='.', label="true")
plt.plot(y_pred, 'r', label="prediction")
plt.ylabel('Value')
plt.xlabel('Time Step')
plt.legend()
plt.show()
```

EXERCISES

19.1 Write a Python program that illustrates the behavior of the discrete Hopfield network as a content-addressable memory using $N = 81$ neurons and the set of handcrafted patterns displayed in Figure 19.12. Display the images of the fundamental memories and the outputted images.

19.2 A schematic of a two-neuron Hopfield RNN is shown in Figure 19.2. Suppose that
$$\frac{dx}{dt} = -x + 7\left(\frac{2}{\pi}\tan^{-1}\left(\frac{\gamma\pi x}{2}\right)\right) + 6\left(\frac{2}{\pi}\tan^{-1}\left(\frac{\gamma\pi y}{2}\right)\right),$$

$$\frac{dy}{dt} = -y + 6\left(\frac{2}{\pi}\tan^{-1}\left(\frac{\gamma\pi x}{2}\right)\right) - 2\left(\frac{2}{\pi}\tan^{-1}\left(\frac{\gamma\pi y}{2}\right)\right).$$

Use Hopfield's theorem to determine a suitable Lyapunov function for this system and plot a surface and contour plot of this function for various γ values of your choice. How many stable states are there in each case?

19.3 The Lorenz system was introduced in 1963 as a simplified model of the weather [12]. This is a famous chaotic system described by the equations:

$$\frac{dx}{dt} = 10(y - x), \quad \frac{dy}{dt} = 28x - y - xz, \quad \frac{dz}{dt} = xy - \frac{8}{3}z.$$

Use this system to create a one-dimensional chaotic time series using either the x, y or z terms, and write a TensorFlow-Keras program LSTM RNN to predict this chaotic time series.

19.4 Download the Google stock price data from Yahoo Finance (or choose your own financial time series data). Alternatively, readers can download the file **Google_Stock_Price.csv** from GitHub. Use Pandas to set up a data frame and plot the financial time series data. Use TensorFlow, Keras, and a LSTM RNN to predict the Google stock price.

Solutions to the Exercises may be found here:

https://drstephenlynch.github.io/webpages/Solutions_Section_3.html.

FURTHER READING

[1] Bao, W., Yue, J. and Rao, Y.L. (2017). A deep learning framework for financial time series using stacked autoencoders and long short-term memory. *PLoS One.* **12** e0180944.

[2] Bianchi, F.M., Maiorino, E., Kampffmeyer, M.C. et al. ((2017). *Recurrent Neural Networks for Short-Term Load Forecasting: An Overview and Comparative Analysis.* Springer, New York.

[3] Fischer, T. and Krauss, C. (2018). Deep learning with long short-term memory networks for financial market predictions. *European Journal of Operational Research.* **270** 654-669.

[4] Garcia, L., Talavan, P.M. and Yanez, J. (2017). Improving the Hopfield model performance when applied to the travelling salesman problem. *Software Computing,* **21**, 3891-3905.

[5] Hochreiter, S. and Schmidhuber, J. (1997). Long short-term memory. *Neural Computation* **99**, 1735–1780.

[6] Hopfield, J.J. (1982). Neural networks and physical systems with emergent collective computational abilities. *Proc. National Academy of Sciences,* **79**, 2554–2558.

[7] Hopfield, J.J. (1984). Neurons with graded response have collective computational properties like those of two-state neurons, *Proc. National Academy of Sciences*, **81**, 3088–3092.

[8] Hopfield, J.J. and Tank, D.W. (1985). Neural computation of decisions in optimization problems. *Biological Cybernetics*, **52**, 141–154.

[9] Korstanje, J. (2021). *Advanced Forecasting with Python: With State-of-the-Art-Models Including LSTMs, Facebook's Prophet, and Amazon's DeepAR*. Apress, New York.

[10] Kostadinov, S. (2022). *Recurrent Neural Networks with Python Quick Start Guide: Sequential Learning and Language Modeling with TensorFlow*. Packt Publishing, Birmingham, UK.

[11] Lawi A. and Kurnia, E. (2021). *Accurately Forecasting Stock Prices using LSTM and GRU Neural Networks: A Deep Learning approach for forecasting stock price time-series data in groups*. LAP Lambert Academic Publishing,

[12] Lorenz, E.N. (1963). Deterministic non-periodic flow. *J. Atmos. Sci.* **20**,130-141.

[13] Michel, A.N., Hou, L. and Liu, D. (2015). *Stability of Dynamical Systems: On the Role of Monotonic and Non-Monotonic Lyapunov Functions (Systems and Control: Foundations and Applications), 2nd Ed.* Springer, New York.

[14] Sagheer, A. and Kotb, M (2019). Time series forecasting of petroleum production using deep LSTM recurrent networks. *Neurocomputing* **323**, 203-213.

[15] Salem, F.M. (2022). *Recurrent Neural Networks: From Simple to Gated Architectures* Springer, New York.

## CHAPTER 20

Convolutional Neural Networks, TensorBoard and Further Reading

Convolutional Neural Networks (CNNs) are covered in the first two sections. The reader is shown how convolution and pooling work in CNNs by means of a simple example. Next, a CNN is used to identify handwritten digits. In Section 3, the reader is introduced to TensorBoard, which is a tool providing performance tables for a Keras-TensorFlow program. The final section briefly discusses the topics Cyber Security, Ethics in AI, the Internet of Things (IoT), Natural Language Processing (NLP) and Reinforcement Learning (RL). A number of references are listed for each topic.

Aims and Objectives:

- To introduce CNNs.

- To introduce TensorBoard.

- To investigate topics beyond the scope of the book.

At the end of the chapter, the reader should be able to

- Convolve an array using a filter (or kernel) array.

- Apply a CNN to pictorial data.

- Run TensorBoard for a Keras-TensorFlow ANN.

- Understand the basics of AI and its applications.

20.1 CONVOLVING AND POOLING

Inspired by the visual system in certain animals, Hubel and Wiesel discovered feature detector cells in the visual cortex, see [8] and [23]. Inspired by the work of Hubel and

DOI: 10.1201/9781003285816-20

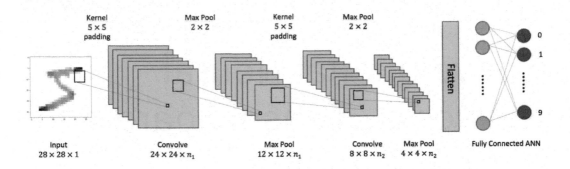

Figure 20.1 A CNN for the MNIST data set of handwritten digits. Reading from left to right, one convolves, then pools, convolves, then pools and finally flattens the data to feed through an ANN in the usual way.

Wiesel, in 1980, Fukushima developed the neocognitron introducing two basic types of layers in CNNs, convolutional and downsampling layers [7]. In the late 1980s, Le-Cun devised LeNet, an early version of a CNN, he used backpropagation ANNs in handwritten digital recognition, see [11] and [12]. The development of graphical processing units (GPUs) in computers in the 2000s greatly improved the performance of CNNs. Real-world applications include handwritten character recognition, face detection, facial emotion recognition, object detection, self-driving autonomous vehicles, language translation, predicting text, X-ray image analysis, and cancer detection, for example.

CNNs are deep learning algorithms, most commonly applied to analyze digital images. They are composed of five types of layer, an input layer, convolutional, pooling, and full connected layers, and an output layer. Figure 20.1 shows a CNN used to recognize handwritten digits, it comprises one input layer, two convolving layers, two pooling layers, and a fully connected ANN layer leading to an output. In order to get a deeper understanding of how the CNN works, Figure 20.2 shows the working of a convolution layer with ReLU activation, a pooling layer, and a flattening layer, so the data can be fed into a fully connected ANN. Looking from left to right, a filter (or kernel) is applied to the input array. The filter in this case is looking for vertical lines. The 5×5 filter slides across the input array one element at a time (the step here is 1) in order to compute the convolved array. Note that the input array has been padded with zeros. This is often used to preserve the original input size. The blue square represents the filter in the top left corner of the input array. Taking element-wise multiplication, leads to the value 2.1, highlighted in blue, in the top left corner of the convolved array. Similarly, the orange box demonstrates the same kind of computation. Next, the ReLU activation function is applied, see Figure 18.2(a), and all negative elements in the array are set to zero (shown in red). A maximum pooling is then applied to this array using a 3×3 filter. The green box in the upper left corner gives a maximum value of 2.1, which can be observed in the top left corner of the pooled array. Similarly, max pooling leads to a value of 1.7 in the bottom right pink box. Finally, the data is flattened, each row is taken in turn to give a 9×1 column vector.

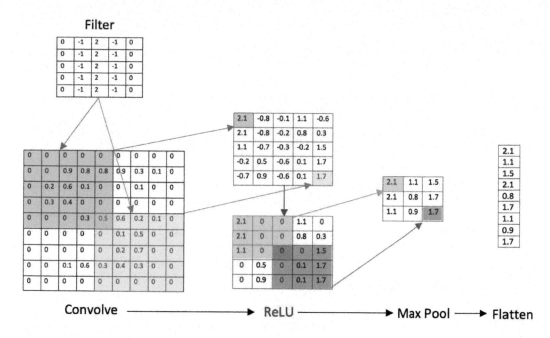

Figure 20.2 Compare with Figure 20.1. A numerical example, where a 5×5 filter (or kernel) slides across the padded input array, one step at a time, to obtain a 5×5 convolved array. One simply uses element-wise multiplication, the blue and orange boxes should help the reader understand. A ReLU activation function is then applied to the convolved array and the negative terms become zero. A 3×3 filter is then applied using a maximum pooling to obtain a pooled 3×3 array, see the green and pink boxes. Finally, the data is flattened to give a 9×1 array that can be fed into the ANN.

Example 20.1.1. Write a simple Python program to perform convolution for the input array and the filter (kernel) displayed in Figure 20.2.

Solution. Program_20a.py lists the program to give the convolved array.

```
# Program_20a.py: Convolve A with K.
import numpy as np
# Input padded array.
A = np.array([[0,0,0,0,0,0,0,0,0],
              [0,0,0.9,0.8,0.8,0.9,0.3,0.1,0],
              [0,0.2,0.6,0.1,0,0,0.1,0.2,0],
              [0,0.3,0.4,0,0,0,0,0,0],
              [0,0,0,0.3,0.5,0.6,0.2,0,0],
              [0,0,0,0,0,0.1,0.5,0,0],
              [0,0,0,0,0,0.2,0.7,0.4,0],
              [0,0,0.1,0.6,0.3,0.4,0.3,0,0],
              [0,0,0,0,0,0,0,0,0]])
```

```
# The filter (kernel), used to find vertical lines.
K = np.array([[0, -1 , 2 , -1 , 0],
              [0 , -1 , 2 , -1 , 0],
              [0 , -1 , 2 , -1 , 0],
              [0 , -1 , 2 , -1 , 0],
              [0 , -1 , 2 , -1 , 0]])
C = np.zeros([5,5])
for i in range(5):
    for j in range(5):
        C[j,i] = np.sum(A[j : 5+j , i : 5+i] * K)
print(C)
```

The next section shows how Keras and TensorFlow can be used to build a CNN to recognize handwritten digits.

20.2   CNN ON THE MNIST DATASET

The MNIST (Modified National Institute of Standards and Technology) dataset consists of 60,000 training images and 10,000 testing images of the 10 digits 0 to 9. The images are all 28×28 pixels in size. More information can be found on the MNIST database homepage:

http://yann.lecun.com/exdb/mnist/

The data can be downloaded through the Keras datasets.

The softmax activation function normalizes input vectors into a probability distribution proportional to the exponentials of those vector values and is often used as the last activation function for CNNs. The softmax activation function is defined by:

$$S\left(x_i\right) = \frac{e^{x_i}}{\sum_{j=1}^{n} e^{x_j}}, \tag{20.1}$$

where x_i are the values in an n-dimensional vector.

Example 20.2.1. Use Keras and TensorFlow to build a CNN which classifies handwritten digits from the MNIST data set.

Solution. Program_20b.ipynb lists the TensorFlow program. The output from the program lists the layers, their output shapes and the number of parameters involved in each layer. The total number of parameters is 253,130, and the accuracy is over 98 percent.

```
# Program_20b.ipynb: CNN network for the MNIST data set.
import tensorflow as tf
import matplotlib.pyplot as plt
```

```
mnist = tf.keras.datasets.mnist # Digits 0-9, 28x28 pixels
(x_train, y_train), (x_test, y_test) = mnist.load_data()
figure(figsize=(6, 6), dpi=80)
plt.imshow(x_train[0])
plt.show()
# Normalize the data.
x_train = tf.keras.utils.normalize(x_train, axis = 1)
x_test = tf.keras.utils.normalize(x_test, axis = 1)
from matplotlib.pyplot import figure
figure(figsize=(6, 6), dpi=80)
plt.imshow(x_train[0],cmap=plt.cm.binary)
plt.show()
# Build the CNN (see Figure 20.1).
input_shape=(28,28,1)
inputs = tf.keras.layers.Input(shape=input_shape)
layer = tf.keras.layers.Conv2D(filters=64, kernel_size=(5,5), \
        strides=(2,2), activation=tf.nn.relu)(inputs)
layer = tf.keras.layers.Conv2D(filters=64, kernel_size=(5,5), \
        strides=(2,2), activation=tf.nn.relu)(layer)
layer = tf.keras.layers.Flatten()(layer)
# The 1st hidden layer with RELU activation.
layer = tf.keras.layers.Dense(128, activation = tf.nn.relu)(layer)
# The 2nd hidden layer with RELU activation.
layer = tf.keras.layers.Dense(128, activation = tf.nn.relu)(layer)
# The number of classifications with softmax activation.
outputs=tf.keras.layers.Dense(10,activation=tf.nn.softmax)(layer)
# Compile the CNN.
model = tf.keras.Model(inputs, outputs)
model.summary()
model.compile(optimizer='adam',
            loss='sparse_categorical_crossentropy',
            metrics=['accuracy'])
model.fit(x_train, y_train, epochs = 3)
x=x_test.reshape((x_test.shape[0],x_test.shape[1],x_test.shape[2],1))
predictions = model.predict([x_test])
print(predictions)
import numpy as np
index = 1000
print(np.argmax(predictions[index]))
plt.imshow(x_test[index].reshape((28,28)))
plt.show()
```

```
Model: "model"
```

Layer (type)	Output Shape	Param #
input_1 (InputLayer)	[(None, 28, 28, 1)]	0
conv2d (Conv2D)	(None, 12, 12, 64)	1664
conv2d_1 (Conv2D)	(None, 4, 4, 64)	102464
flatten (Flatten)	(None, 1024)	0
dense (Dense)	(None, 128)	131200
dense_1 (Dense)	(None, 128)	16512
dense_2 (Dense)	(None, 10)	1290

```
Total params: 253,130
Trainable params: 253,130
Non-trainable params: 0
```

Epoch 1/3: 24s, 12ms/step – loss: 0.1693 – accuracy: 0.9477

Epoch 2/3: 24s, 12ms/step – loss: 0.0584 – accuracy: 0.9822

Epoch 3/3: 24s, 12ms/step – loss: 0.0393 – accuracy: 0.9872

Figure 20.3 Typical output from Program_20b.ipynb.

20.3 TENSORBOARD

TensorBoard is a tool providing visualizations and measurements of your TensorFlow program. It includes scalars, graphs, distributions, histograms, and time series in its outputs. The scalars dashboard tracks scalar values including loss and accuracy metrics, training speed, and learning rate with each epoch. The graphs closely match the Keras model definition in pictorial form. The distributions and histograms give pictorial representations of the weights and the biases and shows their distribution over time. The example below shows how TensorBoard can be used to check the performance of a Keras-TensorFlow program used to classify handwritten digits without using CNNs. Instead, the reader is introduced to Keras dropout layer, where nodes that do not contribute much to the network are removed. In the program below, the dropout rate is set at 0.2, and helps to prevent overfitting.

Example 20.3.1. Run a TensorFlow program of your choice and evaluate its performance using TensorBoard.

Solution.

```
# Program_20c.ipynb: TensorBoard to Evaluate a Networks Performance.
try:
  %tensorflow_version 2.x
except Exception:
  pass
# Load the TensorBoard notebook extension
%load_ext tensorboard
import tensorflow as tf
import datetime
# Clear any logs from previous runs
!rm -rf ./logs/
# Load the MNIST data set.
mnist = tf.keras.datasets.mnist
(x_train, y_train),(x_test, y_test) = mnist.load_data()
# Normalize the data.
x_train, x_test = x_train / 255.0, x_test / 255.0
# Create the ANN and use a dropout layer.
def create_model():
  return tf.keras.models.Sequential([
    tf.keras.layers.Flatten(input_shape=(28, 28)),
    tf.keras.layers.Dense(512, activation='relu'),
    tf.keras.layers.Dropout(0.2),
    tf.keras.layers.Dense(10, activation='softmax')])
model = create_model()
model.compile(optimizer='adam',
              loss='sparse_categorical_crossentropy',
              metrics=['accuracy'])
log_dir="logs/fit/"+datetime.datetime.now().strftime("%Y%m%d-%H%M%S")
tensorboard_callback = tf.keras.callbacks.TensorBoard(log_dir=log_dir, \
  histogram_freq=1)
model.fit(x=x_train, y=y_train, epochs=16, validation_data = \
          (x_test, y_test), callbacks=[tensorboard_callback])
%tensorboard --logdir=logs
```

The graph output in Figure 20.5 gives a clue to what will appear in future editions of TensorFlow. Those readers familiar with MATLAB® may have heard of the Deep Learning Toolbox®. There is a Deep Network Designer app, where users simply click and drag ANN layers to interactively build their own ANN. A simple push of a button then generates the code to run the ANN. A similar feature will undoubtedly appear in future releases of TensorFlow, and eventually ANNs will be used to optimize ANNs [1].

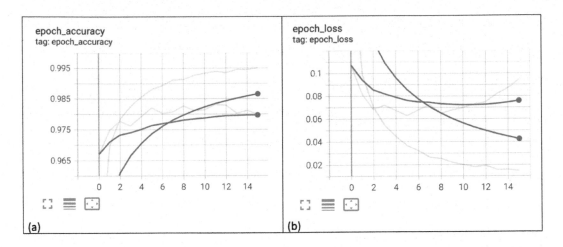

Figure 20.4 Scalar output from Program_20c.py. (a) Epoch accuracy curves for training and validation data. (b) Epoch loss curves for training and validation data.

20.4   FURTHER READING

To end this section, the reader is introduced to some topics which are beyond the scope of this book but follow on nicely from the material presented here.

Cyber Security. AI is being adopted all over the world and is becoming an integral part of business cybersecurity. In 2022, it was estimated that the average business receives 10,000 cyber security alerts every day. Using AI approaches, companies such as CyberArk, Darktrace, FireEye, Google, Microsoft and Rapid7, for example, are already able to detect malware, monitor the spread of misinformation, and stymie phishing campaigns. For further information, please see [9], [15] and [20], for example.

Ethics in AI. There are undoubtedly significant benefits in using AI in cybersecurity and many other arenas of modern life. However, problems of bias, fairness and transparency might arise with these automated systems. Most organizations around the world are now developing codes of ethics. In 2017, more than 100 world leaders developed the 23 Asilomar AI Principles at a conference at Alisomar, California. Research issues, ethics and values, and longer term issues were all addressed. For further information, please see [2], [4] and [6] for example.

IoT. The Internet of Things, otherwise known as IoT, refers to physical devices that are connected over communication networks such as the internet. These interconnected devices have been around since the 1980s and the term IoT was first coined in 1999. With the convergence of multiple technologies including commodity sensors, embedded systems, machine and deep learning, and ubiquitous computing, the future of IoT in business and for the private citizen seems boundless. It is estimated that by 2025, there will be over 21 billion IoT devices worldwide, cities will become "smart," AI will become more prominent, routers will become smarter and be more

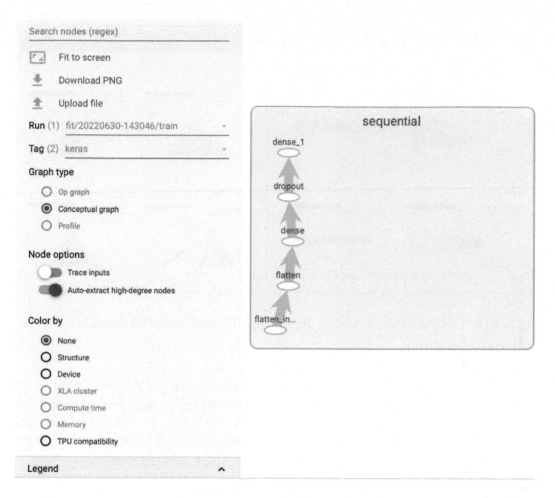

Figure 20.5 Graph output from Program_20c.py. To see a more detailed graph, set the Tag (2) to Default.

secure, fifth generation cellular wireless (5G) networks will boost consumer demand, only autonomous (driverless) cars will be allowed on our roads, and cyber security will become ever more important. For those wanting to pursue a career in IoT, the Raspberry Pi® DIY electronic platform is an excellent place to start. The fourth generation Raspberry Pi 4 was first released in 2019, has up to 8GB random access memory (RAM), a faster quad-core central processing unit (CPU), support for dual displays with 4K resolution, GB ethernet, USB3.0, wireless local area network (LAN), Bluetooth 5 radio, USB-C power, and multiple sensors can be connected simultaneously. Readers are directed to [13], [21] and [25], for example. For books on IoT, see [5], [17] and [24], for example. The author has recently begun some research on IoT [14], investigating long-range wireless connectivity, transmission policy enforcement, and multi-hop routing in quality of service aware IoT networks.

NLP. Natural Language Processing, otherwise known as NLP, is a branch of AI where computers interpret, manipulate and understand human language. There are many

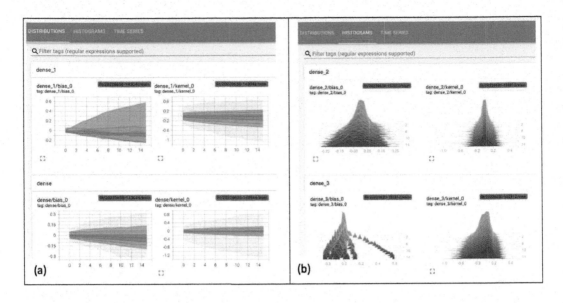

Figure 20.6 Outputs from Program_20c.py. (a) TensorBoard Distributions. (b) TensorBoard Histograms.

applications including chatbots, email classification, sentiment analysis, smart home devices (including Alexa and Google Home), spam detection, text analysis and translation tools, for example. In the Exercises, the reader is asked to run a Google Colab cloud TPU example, where the complete works of William Shakespeare are downloaded and a Keras-TensorFlow NLP LSTM ANN is used to create Shakesperean text based on an initial seed of user text. Granted, the example is not that inspiring at the moment, however, perhaps in 10–20 years time, AI will be able to produce a whole Shakesperean play based on only an outline story. Similarly, AI should be able to create new poetry and even songs from your favorite pop groups. Imagine a brand new song created in the style of the Beatles—or whatever your favorite band happens to be! ChatGPT has revolutionized NLP and natural language generation (NLG) and is a chatbot platform created by OpenAI. Readers may already be using it in their own education and research. For more information on NLP, the reader is directed to [3], [18] and [22].

Reinforcement Learning (RL). The three basic machine learning paradigms are reinforcement, supervised, and unsupervised learning, respectively. RL enables an agent to learn in an active environment based on positive and negative rewards, which are fed back from its own actions and experiences. There are many applications of RL including control theory, game theory, information theory, operational research, simulation-based optimization and swarm intelligence to name a few. RLs mainly utillize Markov decision processes (MDPs) and can be used to play games such as Atari, backgammon, chess and Go. In the future, RL will be used in robotics, and androids (humanoid robots) will discover their own optimal behavior through trial-and-error interactions with their environment. Using high-dimensional tensors, robots will learn to negotiate objects (a video can be thought of as a four or five dimensional tensor, and perception could provide the 6th dimension). As the tensor dimension gets higher and higher, the robot can hear, listen, speak, feel (tactically and emotionally), joke, laugh, cry and who knows where the development will end? The reader may wonder what the tensor rank of a human brain is? For further information on RL the reader is directed to [10], [16] and [19], for example.

EXERCISES

20.1 See Program_20a.py. Apply the filter (kernel) below to the padded array A by hand, and check your hand calculations using this program. The filter is looking for left corner features.

$$K = \begin{pmatrix} -1 & -1 & -1 & -1 & -1 \\ -1 & 2 & 2 & 2 & 2 \\ -1 & 2 & -1 & -1 & -1 \\ -1 & 2 & -1 & 0 & 0 \\ -1 & 2 & -1 & 0 & 0 \end{pmatrix}.$$

20.2 Train a CNN with Keras and TensorFlow on the Fashion MNIST dataset, where the ten classes include T-shirts, trousers, pullovers, dresses, coats, sandals, shirts, sneakers, bags and ankle boots.

20.3 Use TensorBoard on the Boston housing data examples listed in Program_18c.ipynb. Compare the output for a good fit ANN as opposed to an ANN that overfits.

20.4 An example of natural language processing: Shakespeare in five minutes. Run the Google Colab TPU example (written by TensorFlow Hub authors) at the following interactive URL:

```
https://colab.research.google.com/github/tensorflow/tpu/blob/master/
tools/colab/shakespeare_with_tpu_and_keras.ipynb.
```

More examples will appear as Keras and TensorFlow are developed.

Solutions to the Exercises may be found here:

```
https://drstephenlynch.github.io/webpages/Solutions_Section_3.html.
```

FURTHER READING

[1] Abdolrasol, M.G.M., Hussain, S.M.S., Ustun, T.S. et al. (2021). Artificial neural networks based optimization techniques: A review. *Electronics* **10**, 10212689.

[2] Ammanath, B. (2022). *Trustworthy AI: A Business Guide for Navigating Trust and Ethics in AI.* Wiley, New York.

[3] Antic, Z. (2021). *Python Natural Language Processing Cookbook: Over 50 recipes to understand, analyze, and generate text for implementing language processing tasks.* Packt Publishing, Birmingham.

[4] Bartneck, C., Lutge, C., Wagner, A. et al. (2021). *An Introduction to Ethics in Robotics and AI.* Springer, New York.

[5] Dow, C. (2018). *Internet of Things Programming Projects: Build modern IoT solutions with the Raspberry Pi 3 and Python.* Packt Publishing, Birmingham.

[6] Franks, B. (2020). *97 Things About Ethics Everyone in Data Science Should Know: Collective Wisdom from the Experts.* O'Reilly Media, Sebastopol.

[7] Fukushima, K. (1980). Neocognitron: A self-organizing neural network model for a mechanism of pattern recognition unaffected by shift in position. *Biological Cybernetics.* **36**, 193–202.

[8] Hubel, D.H. and Wiesel, T.N. (1959). Receptive fields of single neurones in the cat's striate cortex. *J Physiol.* **148**, 574–591.

[9] Ijlal, T. (2022). *Artificial Intelligence (AI) Governance and Cyber-Security: A beginner's handbook on securing and governing AI systems.* Independently Published.

[10] Lapan, M. (2020). *Deep Reinforcement Learning Hands-On: Apply modern RL methods to practical problems of chatbots, robotics, discrete optimization, web automation, and more, 2nd Ed.* Packt Publishing, Birmingham.

[11] Lecun, Y. (June 1989). Generalization and network design strategies. *Technical Report CRG-TR-89-4.* Department of Computer Science, University of Toronto.

[12] LeCun, Y., Boser, B. Denker, J.S. et al. (December 1989). Backpropagation applied to handwritten zip code recognition. *Neural Computation.* **1**, 541–551.

[13] McManus, S. and Cook, M. (2021). *Raspberry Pi for Dummies, 4th Ed..* For Dummies, John Wiley and Sons, Inc., New York.

[14] Muthana, M.S.A., Muthana A., Rafiq A. et al. (2022). Deep reinforcement learning based transmission policy enforcement and multi-hop routing in QoS aware LoRa IoT networks. *Computer Communications.* **183**, 33–50.

[15] Parisi, A. (2019). *Hands-On Artificial Intelligence for Cybersecurity: Implement Smart AI Systems for Preventing Cyber Attacks and Detecting Threats and Network Anomalies.* Packt Publishing, Birmingham.

[16] Plaat, A. (2022). *Deep Reinforcement Learning.* Springer, New York.

[17] Rossman, J. (2016). *The Amazon Way on IoT: 10 Principles for Every Leader from the World's Leading Internet of Things Strategies.* John E. Rossman.

[18] Sarkar, D. (2019). *Text Analytics with Python: A Practitioner's Guide to Natural Language Processing, 2nd Ed.* Apress, New York.

[19] Sutton, R. and Barto, A.G. (2018). *Reinforcement Learning: An Introduction (Adaptive Computation and Machine Learning series), 2nd Ed.* A Bradford Book, MIT Press.

[20] Tsukerman, E. (2019). *Machine Learning for Cybersecurity Cookbook: Over 80 recipes on how to implement machine learning algorithms for building security systems using Python.* Packt Publishing, Birmingham.

[21] Upton, E.J. (2019). *Raspberry Pi 4 Ultimate Guide: From Beginner to Pro: Everything You Need to Know: Setup, Programming Theory, Techniques, and Awesome Ideas to Build Your Own Projects.* Independently Published.

[22] Vajjala, S., Majumder, B., Gupta, A. et l. (2020). *Practical Natural Language Processing: A Comprehensive Guide to Building Real-World NLP Systems.* O'Reilly Media, Sebastopol.

[23] Hubel, D.H. and Wiesel, T.N. (1962). Receptive fields, binocular interaction and functional architecture in the cat's visual cortex. *J Physiol.* **160**, 106–154.

[24] Wilkins, N. (2019). *Internet of Things: What You Need to Know About IoT, Big Data, Predictive Analytics, Artificial Intelligence, Machine Learning, Cybersecurity, Business Intelligence, Augmented Reality and Our Future.* ASIN, B07PG317XS.

[25] Yamanoor, S. and Yamanoor, S. (2022). *Raspberry Pi Pico DIY Workshop: Build exciting projects in home automation, personal health, gardening, and citizen science.* Packt Publishing, Birmingham.

## CHAPTER 21

Answers and Hints to Exercises

21.1 SECTION 1 SOLUTIONS

Full solutions to the Exercises in Section 1 may be found here:

https://drstephenlynch.github.io/webpages/Solutions_Section_1.html.

CHAPTER 1

1.1 (a) 30.

  (b) 1.6487.

  (c) 0.

  (d) 0.1.

  (e) $\frac{74}{105}$.

1.2 (a) [46 . . . 400].

  (b) $[-9, -5, \ldots, 199]$.

  (c) 6.

  (d) 5.8.

1.3 (a) 45 degrees Centigrade is 113 degrees Fahrenheit, for example.

  (b) Sum of primes when n=100 is, 1060.

  (c) Import the random module.

  (d) [3,4,5] and [9,12,15] are solutions, for example.

  (e) Remember to ignore punctuation marks.

1.4 (a) Each segment is one-fifth the length of the previous segment.

  (b) The motif consists of five segments each of length one third.

  (c) Make sure the turtle points straight up with left turns and right turns.

DOI: 10.1201/9781003285816-21

(d) One segment is replaced with eight segments.

CHAPTER 2

2.1   (a) [45 145 245 345 445 545 645 745 845 945]

    (b) [9 19 29 39 49 59 69 79 89 99]

    (c) [0 1 2 ... 530 540]

    (d) 514

    (e) np.array([[[1,2],[3,4]],[[5,6],[7,8]],[[9,10],[11,12]]])

2.2   (a) Set domain $-5 \leq x \leq 8$.

    (b) Set domain $0 \leq x \leq 2\pi$.

    (c) Set domain $-\pi \leq x \leq \pi$.

    (d) The curve has a point of inflection.

    (e) The coshine curve.

2.3   (a) $(x - y)\left(x^2 + xy + y^2\right)$.

    (b) $[-3 , 10]$.

    (c) $-\frac{2}{7(x+2)} - \frac{1}{4(x-1)} + \frac{15}{28(x-5)}$.

    (d) $\frac{\cos(4x)}{2} + \frac{1}{2}$.

    (e) $x^3 + x^2 y - 3x^2 - xy + 4x - y^2 + 7y - 12$.

2.4   (a) e.

    (b) $12x^3 - 18x^2$.

    (c) $36(2x - 1)$.

    (d) $x^3/3 - x^2 - 3x + c$.

    (e) $e\sin(1)+(x-1)(e\cos(1)+e\sin(1))+...+(x-1)^9 \left(\frac{e\cos(1)}{22680} + \frac{e\sin(1)}{22680}\right)+\text{higher}$ order terms.

2.5   (a) ∞.

    (b) $\{x : 3/2, y : -1/2\}$.

    (c) [(-1/2+sqrt(13)/2,3/2-sqrt(13)/2),(-sqrt(13)/2-1/2,3/2+sqrt(13)/2)].

    (d) 610.

    (e) 10460353200.

2.6   (a) $\begin{pmatrix} -8 & 11 & 7 \\ -10 & 9 & 11 \\ -2 & 17 & 13 \end{pmatrix}$.

    (b) $\begin{pmatrix} 7 & -3 & 1 \\ 8 & -2 & -1 \\ 15 & -7 & 3 \end{pmatrix}$.

(c) $\begin{pmatrix} -5/9 & -2/9 & 4/9 \\ -1/9 & 5/9 & -1/9 \\ 1/6 & -1/3 & 1/6 \end{pmatrix}$.

(d) $\begin{pmatrix} 771905 \\ 575870 \\ 1467293 \end{pmatrix}$.

(e) $\{0 : 1, 1 : 2\}$. The eigenvalue 1 has multiplicity 2.

2.7 (a) -22-42j.

(b) 32.2025.

(c) -23.3738-611.4766j.

(d) 22.9791-14.7448j.

(e) $7.8102e^{0.8761j}$.

CHAPTER 3

3.1 (a) $x(t) = \frac{C1}{t}$.

(b) $x(t) = C2e^{\frac{1}{t}}$.

(c) $x(t) = -\frac{1}{\tanh(C3-t)}$.

(d) y(t)=C4*exp(t*(-5-sqrt(21))/2)+C5*exp(t*(-5+sqrt(21))/2)-2*cos(t).

3.2 Set the domain and range to $-5 \le x, y \le 5$.

3.3 Set the domain and range to $-1.2 \le x, y \le 1.2$.

3.4 Set the domain and range to $-4 \le x \le 4$, and $-5 \le y \le 5$.

3.5 Set the range to $-3 \le y \le 3$.

CHAPTER 4

4.1 (a) $161 - 72\sqrt{5}$.

(b) $[-1/2, 2]$.

(c) $[(-1/2+sqrt(3)/2,1/2+sqrt(3)/2), (-sqrt(3)/2-1/2,1/2-sqrt(3)/2)]$

(d) $y = -2x + 10$ meets the circle at $(3, 4)$.

(e) Set the domain to $-2 \le t \le 2$.

4.2 (a) $(x - 3)(x + 1)(x^3 - 3)$.

(b) Set the domain to $-6 \le x \le 6$.

(c) $-32x^5 + 80x^4 - 80x^3 + 40x^2 - 10x + 1$.

(d) $\frac{dy}{dx} = 32x^7 - 6x, \quad \frac{d^2y}{dx^2} = 2\left(112x^6 - 3\right)$.

(e) 4.1569.

4.3 (a) $[4, -39]$

(b) $[(\log(2/25)/2 + \text{j*pi})/\log(5), -1 + \log(2)/(2\text{*}\log(5))]$. One root is complex.

(c) Use random.sample.

(d) Mean_data1 $= 32.1333$; median_data1 $= 34$, mode_data1 $= 34$, var_data1 $= 512.2667$, stdev_data1 $= 22.6333$.

(e) In Venn3, use subsets$=(4,3,5,7,0,0,0)$.

4.4 (a) Mean $= 7.5$, variance $= 5.25$, standard Deviation $= 2.2913$.

(b) $H_0 : p = 0.6, H_1 : p < 0.6$. Do not reject H_0. Insufficient evidence to suggest Thalia is correct.

(c) Use plt.plot(xs , ys).

(d) $\{a : -3, b : -4\}$.

(e) Use plt.plot(pts).

(f) Use proof by mathematical induction.

CHAPTER 5

5.1 (a) Use the series command.

(b) 717897987691852588770248.

(c) $g(f(-1)) = 2$.

(d) $-\frac{2}{49}$.

(e) 0.6435.

5.2 (a) $\frac{10}{7(2x-5)} - \frac{3}{14(x+1)} - \frac{1}{2(x-1)}$.

(b) Curves are the same.

(c) $\frac{xe^{\sin(x)}\sin(x)}{\cos^2(x)} + xe^{\sin(x)} + \frac{e^{\sin(x)}}{\cos(x)}$.

(d) $-x^3\cos(x) + 3x^2\sin(x) + 6x\cos(x) - 6\sin(x) + c$.

(e) An ellipse.

5.3 (a) 31.6386.

(b) $N(8) = 0.2021$ mg.

(c) $x(4) = 2.2935174$.

(d) $k = 4/255$; $P(1 < X < 2) = \frac{1}{17}$.

(e) Integrate the PDF to get the CDF.

(f) $H_0 : \mu = 80; H_1 : \mu > 80$. Use the z-test, $z = 2 > 1.645$. Reject H_0, and the managers claim is supported.

5.4 (a) Distance$=10$m; time$=1.4286$ s.

(b) $F_R = 6.40312$ N.

(c) Forces $R = 333.2$ N and $T = 352.8$ N.

(d) Initial velocity in y-direction is 26.5 m/s.

(e) $\mu = \frac{1}{\sqrt{3}}$. Acceleration$=1.9650 \ ms^{-2}$.

21.2   SECTION 2 SOLUTIONS

Full solutions to the Exercises in Section 2 may be found here:

`https://drstephenlynch.github.io/webpages/Solutions_Section_2.html`.

CHAPTER 6

6.1 (i) $a = 0.2$, Fixed points at $c_1 = 0$, stable; $c_2 = 0.1553$, unstable; and $c_3 = 0.9455$, stable. (ii) $a = 0.3$, Fixed points at $c_1 = 0$, stable; $c_2 = 0.1704$, unstable; and $c_3 = 0.8972$, unstable.

6.2 There are three critical points in the first quadrant. The origin is a saddle point, the critical point at $(4, 0)$ is a saddle point, and the critical point at $(5/4, 11/4)$ is a stable focus. The species co-exist.

6.3 $\frac{dS}{dt} = -\beta S (I_1 + I_2 + \xi (A_1 + A_2) + \eta L_2)$, $\frac{dL_1}{dt} = \beta S (I_1 + I_2 + \xi (A_1 + A_2) + \eta L_2) - \epsilon L_1$, $\frac{dL_2}{dt} = \epsilon (L_1 - L_2)$, $\frac{dI_1}{dt} = (1 - \pi)\epsilon L_2 - \gamma I_1$, $\frac{dI_2}{dt} = \gamma (I_1 - I_2)$, $\frac{dA_1}{dt} = \pi \epsilon L_2 - \gamma A_1$, $\frac{dA_2}{dt} = \gamma (A_1 - A_2)$, $\frac{dR}{dt} = \gamma (I_1 + A_2)$. Maximum of $A_1 = 6604$. Maximum of $I_1 = 59440$.

6.4 Without holding, a convex clockwise hysteresis loop is formed. See the "Hysteresis in Muscle" paper cited in the chapter.

CHAPTER 7

7.1 Balanced equation is:

$$3NaHCO_3 + H_3C_6H_5O_7 \rightleftharpoons Na_3C_6H_5O_7 + 3H_2O + 3CO_2.$$

7.2 $|x| \to 1$ and $|y| \to 0$.

7.3 $|O| = 46805158.11585307$ molec/cm cubed, $|O2| = 6.999019071196589e + 16$ molec/cm cubed, $|O3| = 6539509754323.823$ molec/cm cubed.

7.4 At a KCl concentration of 0.1 AgCl solubility is 3.24E-06. See Figure 7.4.

CHAPTER 8

8.1 Use df.set_index("Date") to set the Date column as an index.

8.2 (a) Shade the area above the lines. The vertex at $(25, 50)$ is the point where the cost function is minimal. (b) $z_{max} = 268$, and $x_1 = 1.8, x_2 = 20.8, x_3 = 1.6$.

8.3 There are three clusters and more outliers. The OCSVM has far more sets.

8.4 From the root, there are 2, 4 and 6 branches. In the root box, hdlngth<= 91.55, squared_error=4.128, samples=70 and value=4.014.

CHAPTER 9

9.1 Plot the graph in Spyder to find intersection points. The gradient of the expansion path line is 0.06.

9.2 (i) The solutions are steady states. (ii) The solutions oscillate.

9.3 The efficient frontier curve is a hyperbola.

9.4 The code is:

```
!pip install py_vollib_vectorized
from py_vollib.black_scholes import black_scholes as bs
from py_vollib.black_scholes.greeks.analytical import \
      delta,gamma,vega,theta,rho
r , S , K , T , sigma = 0.01 , 30 , 40, 240/365 , 0.3
print('BlackScholes',bs('c',S,K,T,r,sigma))
print('delta',delta(type,S,K,T,r,sigma))
print('gamma',gamma(type,S,K,T,r,sigma))
print('vega',vega(type,S,K,T,r,sigma))
print('theta',theta('c',S,K,T,r,sigma))
print('rho',rho('p',S,K,T,r,sigma))
```

CHAPTER 10

10.1 $i(t) = 5\sqrt{7}\sin(\sqrt{7}t/2)/14 - 5\cos(\sqrt{7}t/2)e^{-t/2} + 5e^{-t}/2$.

10.2 Define a function for the ODEs as:

```
def Coupled_LC(X, t, alpha , omega):
    x1, y1, x2, y2 = X
    dx1 = y1
    dy1 = -omega**2*((1+alpha)*x1+alpha*x2)
    dx2 = y2
    dy2 = -omega**2*((1+alpha)*x2+alpha*x1)
    return(dx1, dy1, dx2 , dy2)
```

10.3 Define a function for the ODEs as:

```
def Mass_Spring(X, t, k1 , k2 , k3 , k4):
    x1 , y1 , x2 , y2 , x3 , y3 = X
    dx1 = y1
    dy1 = (-k1*x1+k2*(x2-x1))/m1
    dx2 = y2
    dy2 = (-k2*(x2-x1)+k3*(x3-x2))/m2
    dx3 = y3
    dy3 = (-k3*(x3-x2)-k4*x3)/m3
    return(dx1 , dy1 , dx2 , dy2 , dx3 , dy3)
```

10.4 (i) Quasiperiodic behavior. (ii) Chaotic behavior—sensitivity to initial conditions.

CHAPTER 11

11.1 The area of the true fractal is 2 units squared. Use an infinite geometric progression in your proof.

11.2 The box-counting (fractal) dimension of Barnsley's fern shown in Figure 10.3 is approximately 1.7807. It is more dense than the Sierpinski triangle.

11.3 The fractal dimension D_f is approximately 1.462. The f-α curve is skewed left.

11.4 $\alpha_{max} = 3.195$, $\alpha_{min} = 1.101$ and $D_0 = 2$.

CHAPTER 12

12.1 Take $k = 6$, you should obtain a 712×712 image.

12.2 There are 77,498 yellow pixels.

12.3 I used the **skeletonize** library. The vascular architecture is 14,234 pixels long.

12.4 Percentage area of brain tumor is approximately 11.27%.

CHAPTER 13

13.1 Backward Euler, $y_{n+1} = y_n + hf(x_{n+1}, y_{n+1})$. Use Newton-Raphson to find roots.

13.2 Adams-Bashforth, in the code, set K2=K1, K1=f(t[i],y[i]), and y[i+1]=y[i]+h*(3*K1-K2)/2.

13.3 Advection, in the vectorized scheme, in the code, U[1:-1] -= C * (U[1:-1] - U[0:-2]).

13.4 Heat diffusion in plate, vectorized code, Unew[1:-1,1:-1]=U[1:-1,1:-1]+alpha*dt/dx2*(U[:-2,1:-1]-2*U[1:-1,1:-1]+U[2:,1:-1])+alpha*dt/dy2*(U[1:-1,:-2]-2*U[1:-1,1:-1]+U[1:-1,2:]).

CHAPTER 14

14.1 (i) $\beta = 0.05$, period-4 behavior, the power spectrum shows a single spike; (ii) $\beta = 0.12$, quasiperiodic behavior, the iterative points appear to form a closed curve and , the power spectrum shows a discrete spikes; (iii) $\beta = 0.3$, chaos, a broad power spectrum. The system is sensitive to initial conditions.

14.2 Splitting into real and imaginary parts, define:

```
def ikeda(X):
    x, y = X
    xn = A + B*x*np.cos(x**2 + y**2) - B*y*np.sin(x**2 + y**2)
    yn = B*x*np.sin(x**2 + y**2) + B*y*np.cos(x**2 + y**2)
    return (xn, yn)
```

The animation begins with a fixed point of period one. This moves and jumps to another stable fixed point of period one (hysteresis cycle). There is then period doubling and undoubling in and out of chaos.

14.3 The ODE is defined by the function:

```
def xdot(X,t):
    x , y = X[0] , X[1]
    dxdt , dydt = y , kappa - 0.6 * y - np.sin(x)
    return [dxdt , dydt]
```

See Figure 13.5, ramp κ up and down.

14.4 A chaotic orbit is observed. You have to take a very small step size.

CHAPTER 15

15.1 Between 1982–2000, the gradient of the linear regression curve was positive, indicating that the carbon dioxide emissions per capita were increasing with gross domestic product per capita.

15.2 Steady state after 11 iterations is $[p_1, p_2, p_3] = [0.62 \quad 0.33 \quad 0.07]$.

15.3 F statistic is 3.295, the p-value is $0.0188 < 0.05$, meaning there is a statistically significant difference (we reject H0). We have strong statistical evidence that the treatment has an effect on NO levels, there is a significant difference in mean concentrations due to drug treatments (we accept H1).

15.4 On average, the gambler would expect to lose half of the bankroll after 1850 spins.

21.3   SECTION 3 SOLUTIONS

Full solutions to the Exercises in Section 3 may be found here:

https://drstephenlynch.github.io/webpages/Solutions_Section_3.html.

CHAPTER 16

16.1 The neuron is a threshold oscillator. As the input current, i, increases, the stable fixed point turns unstable and a stable limit cycle bifurcates, the amplitude grows to a maximum, and then shrinks back down to zero, where a stable fixed point emerges once again. Inside the stable limit cycle, there is an unstable critical point.

16.2 Two limit cycles are stable and one is unstable. As in Exercise 16.1, the reader could produce an animation as the input current increases and decreases. In this case, there is a bifurcation from one limit cycle to another.

16.3 For the full-adder, take $w_1 = 0.8$, $w_2 = 0.45$ and $x_1 = 1.5$. See the paper entitled "Oscillatory threshold logic," cited in the chapter, for more information.

16.4 Couple four oscillators and set:

```
# Input current.
def Input_1(t):
    return 1*(t>1000)-1*(t>1010)
def Input_2(t):
    return 1*(t>2020)-1*(t>2030)
```

CHAPTER 17

17.1 XOR(0,0)= 9.357622968839299e-14, XOR(0,1)= 0.9999999999999065, XOR(1,0)= 0.9999999999999065, XOR(1,1)= 9.357622968891759e-14.

17.2 Keep the biases constant in this exercise, y = 0.28729994077761756, w13_new = 0.1521522763401024, w23_new = 0.10215227634010242, w11_new = 0.20020766275648338, w12_new = 0.15013942100503588, w21_new = 0.2502076627564834, w22_new = 0.3001394210050359

17.3 Only two neurons are required in the hidden layer.

17.4 Compare the bifurcation diagrams with the stability diagram shown in Figure 17.7(b). Note that the instability regions could be chaotic, or periodic regions of period greater than one.

CHAPTER 18

18.1 Edit Program_18b.py.

18.2 Use the piecewise function from the numpy library to plot the ReLU and leaky ReLU functions. The derivatives of the activation functions are used in the backpropagation algorithm.

18.3 The accuracy is nearly always one when the Adam learning rate is 0.1 and 500 epochs are applied.

18.4 Read the Keras web pages and run examples.

CHAPTER 19

19.1 Run the program a number of times with the non-fundamental memories and you should eventually discover spurious steady states.

19.2 The Lyapunov function is:

$$V(a_1, a_2) = -\frac{1}{2}\left(7a_1^2 + 12a_1a_2 - 2a_2^2\right) - \frac{4}{\gamma\pi^2}\left(\log\left(\cos(\pi a_1/2)\right) + \log\left(\cos(\pi a_2/2)\right)\right).$$

Plot contour and surface plots for varying γ of your choice. There are two local minima when $\gamma = 0.5$. You can rotate the surface in Spyder.

19.3 The LSTM RNN gives an excellent approximation. I used variable x in my program.

19.4 Google stock price data is found here (last accessed July 2022):

https://finance.yahoo.com/quote/GOOG/.

CHAPTER 20

20.1 The convolved array is:

```
[[ 4.9  7.9  3.9  3.1  2. ]
 [-0.5 -2.  -4.2 -2.4 -1.3]
 [-0.3 -1.8 -1.9 -0.8 -0.4]
 [ 0.8  1.6  3.   1.1  0.7]
 [-0.9 -1.8 -0.5 -2.1 -0.6]]
```

20.2 A working interactive program can be found here:

https://colab.research.google.com/github/tensorflow/tpu/blob/master/tools/colab/fashion_mnist.ipynb.

20.3 See the overfitting curves in Figure 18.4.

20.4 The URLs are given in the question, the Google notebooks are interactive, simply run the cells.

Index